FUTURE

How Adaptive Leaders Anticipate Change, Decode Signals,

and Build What Comes Next

Christina Diane Warner

Published in the United States of America
ISBN 978-1-7331496-8-6

www.christinadianewarner.com

First Printing Edition, 2025

INTRODUCTION

Marketing has always been a series of "what if" questions. So, here's mine. What if the future isn't waiting for us up ahead? What if it's already playing out right now?

Think of it like a stream. Every droplet upstream eventually joins the ocean, but by the time it gets there, it's already been in motion for ages. That's how trends work. By the time the world calls them "mainstream," they've already been rippling under the surface.

This book is my way of taking you upstream so that you can see the trends before they become trendy.

Every day, I talk to founders and creatives who can feel that things aren't the way they used to be. Data is moving at inhuman speeds and markets change direction before the quarterly report even gets to their desks. We're feeding algorithms and they're feeding us right back, which means consumer expectations are updating faster than your iOS.

Trying to get back to what you might perceive as normal is a losing game because the pace of change won't slow down for your planning cycle. Our old strategies, which were built on prediction and reaction, don't stand a chance anymore. Yesterday's data can't predict tomorrow's reality because proof comes too late now. By the time something's validated, someone else has already moved on it.

The future doesn't reward prediction like it once did. To be on the winning end of a trend, you're going to need to spot the changes before they become obvious. Then, you'll need to act while everyone else is still waiting for certainty. Try to remember the stream analogy throughout this book because every disruption starts as somewhat of an outlier.

The FUTURE framework will take you upstream to find those outliers. It'll help you:

- Find signals
- Understand Meaning
- Transform Assumptions
- Unlock New Patterns
- Render Scenarios
- Execute & Embed

Table of Contents

Introduction iii

FUTURE

Part I: Find Signals

Chapter 1 – The Edges Where the Future Begins 11

    The Neuroscience of Attention 13

    Pattern Recognition & Future Perception 15

    Difference Between Loud Nothingness & Signal 16

Chapter 2 – Tools of Signal Intelligence 19

    Edge Network Behavior 22

    Memetics & Signal Propagation 24

    Startup Ecosystems & Venture Signals.. 26

    Building a Signal Radar 28

Chapter 3 – The Emotional Pulse of a Signal 31

    Emotional Feedback Systems.. 32

    When Signals Become Self-Expression 34

    Invisible Operating System 36

    Desire Mapping 38

    Perception Extension with AI 39

Part II: Understand Meaning

Chapter 4 – Turning Data into Meaning 43

Use STEEP Framework for Meaning-Making 46

"Jobs-to-Be-Done" for Human Motivation 51

Pattern Framing & Sensemaking 52

Chapter 5 – Insight Architecture 56

Mapping Meaning 57

Minimum Viable Bets... 60

Narrative Prototyping 63

Operationalizing the Insight.. 64

Part III: Transform Assumptions

Chapter 6 – The Neuroscience of Unlearning 67

Cognitive Rigidity and Its Effects 70

Expertise as a Barrier to Adaptation 72

The Adaptive Mindset 73

Chapter 7 – Bias, Blindness, and Breakthrough 77

Your Strategic Blind Spots 78

Inversion Strategy 81

Reverse Mentorship 84

Part IV: Unlock New Patterns

Chapter 8 – The Mechanics of Momentum 91

The S-Curve of Adoption 92

Acceleration Dynamics 95

Stacking & Convergence 99

AI-Assisted Momentum Detection 101

*Part V: **R**ender Scenarios*

Chapter 9 – Designing Alternate Futures 105

Thinking in Possibilities Instead of Probabilities 106

Introduction to Scenario Mapping 109

Using the Futures Triangle 112

Using the Futures Wheel 114

Turning Strategic Imagination into Actions 116

Chapter 10 – Consequences, Risks, and Openings 119

Second-Order Thinking 120

Category Inversion & Industry Disruption 121

*Part VI: **E**xecute & Embed*

Chapter 11 – Building a Future-Ready System 124

Scanning Cadences 125

Foresight Pods & Cross-Functional Insight Teams 127

Dashboards & Visibility Systems 129

Chapter 12 – The Future-Ready Leader 132

Leading with Long-Term Vision 133

The Leader as a Perceptor 138

Going From Operator to Architect 141

In Closing 143
References 145

CHRISTINA DIANE WARNER

FUTURE

How Adaptive Leaders Anticipate Change, Decode Signals,

and Build What Comes Next

Part I: **F**ind Signals

(Notice)

Chapter 1 –
The Edges Where the Future Begins

I want to talk about those edges upstream because I have seen them throughout my career across technology, consumer behavior, culture, and marketing. The future first shows signs of its existence in places that seem too small or too niche to have an impact. That's the point. They're not commercialized yet and they're not constrained by expectations, which means they operate with pure creative energy from a place of honest desire. Market leaders are the ones who've been paying attention before anyone else was watching.

This all feels a little like smoke and mirrors, so let me explain a little further what I mean by "edges" and "upstream." I'm talking about fringe communities. What we need to understand is that they're not fringe because they're unimportant. They're fringe (or on the fringe) because traditional institutions haven't come up with ways to categorize them yet. Early adopters gather there because they are motivated by identity and exploration rather than safety or validation. These are the people who are willing to test and challenge what exists. They don't ask for permission. Inventors rarely do. If you look at the origins of almost any major market change, you'll see that exact pattern. The personal computer was once a hobbyist experiment. Streetwear was once a small cultural code that was shared between skaters and graffiti artists. Wellness culture started as a movement of people who were desperately looking for an alternative to healing when traditional medicine failed them. The point is that what starts out as a fringe interest normally becomes a cultural reset.

The reason fringe communities drive early adoption is simple. It's because they operate outside the constraints of mass expectation. They don't have to protect an existing model or please a wide audience. They go after what feels more relevant and truthful to them, which is the type of freedom that allows new behavior to come through. By the time larger organizations take notice of this, that underlying change is already activated and moving. If you don't believe me, just ask the tech giants who were dismissive of cryptocurrency until just a few years ago. You look at record labels that resisted indie creators until streaming platforms came along. Traditional institutions and organizations are almost always the last to see what's coming because they're so deeply focused on defending the present instead of exploring the future.

We're seeing it now with the resistance to AI, but I'm not going to go down that rabbit hole just yet.

Essentially, what I'm getting at is that subcultures act like early laboratories for the next wave of human desire. Underground creators test new aesthetics and tools long before the market validates them. These creators are expressing a need that hasn't been met (yet). When you see something gaining momentum in a subculture, you're looking at a signal of deeper desire moving through a population before it becomes socially acceptable. The adoption pathway goes from subculture to early adopter to niche mainstream to mass market more often than not.

Now, what's there to all of this?

Well, for starters, understanding the difference between a fringe signal and a mainstream trend is going to help you stay ahead of the curve. A mainstream trend is visible and it's already monetized. Is it possible to monetize right along with everyone else? Sure. Why not? The only problem is that, at that stage, your only options are reaction and competition. A fringe signal, on the other hand, is still in bloom, so you've got time before it reaches that scale. You've got leverage because it's closer to the origin point of change and timing is everything.

Being able to build from a signal lets you move with the momentum of the stream instead of swimming against it.

The Neuroscience of Attention

Alright, fish and stream analogies aside, now that we know how important the fringe is, the real risk is missing the early signs of change that most people never see. The good news is that those signs aren't invisible. They're just filtered out by the brain before they reach conscious awareness. That isn't necessarily because people are careless but just because the brain is designed for efficiency. Our minds like the path of least resistance, so we tend to go for what's obvious.

The human brain processes an insane amount of information every second, and almost all of it is filtered out automatically by the Reticular Activating System (RAS). If we tried to hold onto all of that, it would most likely lead to a meltdown. The brain has to decide what's useful and what can be ignored and it does this by matching new information against familiar patterns. If it recognizes something, it allows it through. If it doesn't recognize it, it often categorizes it as something that's irrelevant. This is actually helpful when you're trying to get through daily life and not be overwhelmed. That said, it's unhelpful when the world is changing and you need to spot something new.

So, let's talk about the attention your brain gives to certain things. One of the most powerful drivers of attention is novelty, but novelty only matters if it carries some emotional weight with it. That's because the brain isn't looking solely for what's new. It's looking for what's meaningful. If something is new and sparks excitement or discomfort, the brain flags it and pays attention.

You're not consciously doing this, so you need to consciously undo it. If the information doesn't matter to your survival or your identity, the signal never reaches conscious awareness. It doesn't even enter your field of awareness. If your brain is still tuned to past priorities, you won't see it coming.

What this means for leaders is simple. You can't rely on instinct alone unless that instinct has been trained in new environments. Instinct based on past success will work against you in times of rapid change. Your brain has to be rewired to notice anomalies and you do this by intentionally exposing yourself to new spaces and forcing your brain to engage with unfamiliar signals. Every time you notice something that challenges your assumptions, focus on it because it will help you increase your perceptual range.

The brain isn't a fixed system, which means you can train it to become more adaptive and more perceptive.

Pattern Recognition & Future Perception

If you can train your brain to become more perceptive, you can enhance your pattern recognition skills and forecast more accurately. Now, that sentence might be reinforcing something that it shouldn't and I'll explain what I mean by that. When people hear the word forecasting, they think it means trying to predict what will happen next, but remember that it's that mindset which keeps them stuck. We're trying to recognize the patterns that are already forming, yes, but for what? The answer is to understand what direction they're pointing in. Prediction is tied to reaction and a trend. Pattern recognition is tied to positioning and the pattern itself. A trend is the result. A pattern is the beginning.

As leaders, we can take advantage of our sensitivity to human desire and cultural movement. That perceptual sensitivity gives us the ability to notice small changes in tone, language, search behavior, content creation, emotional expression, and product experimentation. Well, in theory, anyway. You don't just become a marketer and inherently develop that skillset. It's not a talent you're born with either. Like any skill, it's something that you can develop with enough time and effort. The more time you spend studying the edges of culture and those emerging behaviors, the more quickly your brain will learn to pick up on patterns.

To build this skill, you need a structure in place. Anything worth doing is worth doing well or properly. Try to track signals using a simple pattern catalog that covers what you're noticing, where you saw it, what emotional driver it points to, and whether you're seeing similar signals in other domains. If you're not sure of what to track exactly, I'll unpack that for you a little later.

When I track, I'm not looking for proof. I'm looking for recurrence. If I see a change in how people talk about productivity, then I see a startup using that same language, then I see a micro-influencer building a following around it, I know it's a pattern forming across categories.

It takes discipline to build that skill, but that doesn't make it impossible. That sets you up so that you can spot the difference between loud nothingness (or noise) and a signal. In Part VI, you're going to have a couple of practical exercises for that very purpose.

Difference Between Loud Nothingness & Signal

One of the most important skills I've developed in my career is learning how to separate noise from signal. Tracking patterns means nothing if I can decipher between the two. I'd just be recording absolute nonsense if I couldn't discern in that way. What I've noticed is that once people are able to climb upstream, they think that's it. They can now implement tactics based on what they've seen on the fringes, but they can't. In fact, they're probably in the same position as the people who didn't make the trek in the first place.

The reason why most of the people who've started finding places to record patterns still struggle to understand what's coming next is not because they do not have access to information. It's because they treat all information as equal. It's not. Some information is just noise. It might look important for a second, but it has no lasting direction. You have to know how to stop chasing what's temporary and start aligning with what's inevitable.

One thing that noise thankfully has going for it is that it's easy to recognize if you know what to look for. It is usually fast, viral, everywhere, and superficial. If signals are novelty plus emotion, noise is novelty without any depth. I say that it's everywhere because it shows up in every corner of the media and more. It triggers curiosity or outrage, but it fades just as fast because it doesn't connect to any long-term desire or structural change. A meme that explodes and disappears within weeks is noise. A product that goes viral thanks to a single influencer moment, without repeated adoption across different groups, is noise. Noise is reactive because it responds to attention instead of responding to underlying demand.

The difference between that and a signal is that signals typically (not always, but most of the time) start quietly in a small group or niche environment. It's quite rhythmic in how it repeats across different contexts. The emotional energy grows and it creates new language. It's consistent. It also ticks a couple of boxes.

☐	Tied to a core human driver (identity, belonging, control, health, security, status)
☐	Appearing independently in multiple communities (not the result of a viral spike or media push)
☐	Showing signs of adoption rather than conversation (people are using it, not just talking about it)
☐	Creating real behavior change (shifts actions, habits, or decisions)
☐	Moving forward without institutional promotion (the momentum is organic, not manufactured)

Table 1: Signal Surety Checklist

A viral product trend might spike in sales for a brief moment, but a cultural shift changes how people define value. When I saw the early signals of direct-to-consumer brands, it wasn't because they were popular. The simple fact was that they represented a cultural shift toward transparency and personal agency. There was a connection to brand values. That was a signal (or a sign) of a new expectation.

To sum up, signal-to-noise ratio (SNR) is basically the filter I use to separate useful insight from background chatter. I treat it like instrumentation by making it a quick scoring system for every signal I log. Each input gets a score across five dimensions.

1. Desire Link
2. Cross-Domain Recurrence
3. Behavior-Change Evidence
4. Velocity Slope
5. Independence

Each is rated from zero to two and then I add them up. Anything that scores seven or higher triggers action. Five or six goes on the watchlist. Four or less gets archived. You ignore volume and pay attention to the rate of meaning.

The only way to build the ability to tell the difference is through consistent exposure to edge environments. The more you sit and actively observe where ideas are born and how they evolve, the easier it becomes to recognize which ones have depth.

Chapter 2 –
Tools of Signal Intelligence

When I scan for what's coming next, I'm not guessing and I'm not looking for predictions. I'm actually scanning the horizon. It might sound a little spacey or woo-woo, but that's exactly what I'm doing. If you aren't aware, horizon scanning is the practice of looking across different fields and industries to find the earliest signs of change before they become visible in mainstream channels. Community watching is also a part of this. Now, what's interesting is that, in the past, this all required endless manual effort and a wide network of inputs. Today, artificial intelligence has provided us with the ability to extend our field of perception. I know that there is so much talk about AI taking jobs (including marketing jobs) and it's made everyone run red with anger at a tool.

But I need to stress that it is a tool.

The people doing all this replacing are at fault here. So, we need to get our emotions out of it and see AI for what it can do for us while we're still gainfully employed (or self-employed). I'm not going to argue semantics about super-intelligent AGI because we're not there yet, and we'll know for sure when we are. Believe me.

My point is that AI doesn't replace human foresight. It increases our range and speed. It's also way more adept at pattern awareness, which means I can spend less time searching and more time interpreting.

That's simply because AI tools that are designed for horizon scanning operate across multiple data streams at once. Some tools focus on search intent and keyword evolution, which shows me what questions people are asking in large numbers. Others track content velocity and sentiment shifts. They also look at how ideas spread from niche forums to broader social channels. There are tools that scan patent databases and funding announcements. We can get information on early product launches to find emerging tech direction. Then, you've got AI systems that analyze language itself and can actually pick up on these subtle changes in emotional tone or the introduction of new concepts that didn't exist in public conversations six months earlier. All of these signals would definitely seem like they're isolated events if you look at them on their own, but when you synthesize them together, you can see directional movement.

So, how does AI do this? Well, for many of the tools that we have out there right now, they're programmed to pick up on emerging topics by analyzing weak signals at scale. AI essentially taps into rising frequency, so an emotionally charged keyword that starts appearing in new contexts is often the first ping on its radar. That ultimately points to a shift in perception. For example, if I see a word that starts getting traction in wellness communities and then shows up in productivity conversations, then in startup culture, I know it is gaining momentum in multiple dimensions. AI can spot that directional spread in real time across platforms. It monitors anomalies and can predict or, at the very least, quickly see these unexpected spikes. Then, it alerts me to the emergence of new patterns. This is super important because early change doesn't begin with volume but with variance.

I don't see the value of AI in horizon scanning in its ability to collect more data, which we already know it's incredibly successful at doing. I'm more focused on its ability to filter through that data. Our biggest challenge today isn't a lack of information. In fact, we're heavily oversaturated, so much so that we're practically drowning in information. So, I want AI to sift and filter, but AI can't really tell me what something means from a human perspective. That's still my role. So, AI gives me the raw signal and I decide whether it points to a meaningful change.

To get real value out of any of this, I'm measuring how much meaning AI can bring to the surface for me to decide on. I don't care about the size of a spike or how many mentions a keyword has yet. I'm more concerned with the emotions sitting behind it. AI can process patterns faster than I ever could, but it can't tell me why people care. Never forget that this is where human judgment comes in. My job is to translate patterns into perception, which basically means to see how those signals connect to what people want.

Horizon scanning is the next logical step in learning how to see change before it becomes obvious.

Edge Network Behavior

So, we know that the edges are where we find signals, but how do we know where the edges are?

The secret is that you already do.

When I talk about edge networks as I've been doing, I'm not describing a demographic. I'm describing an ecosystem, which (in a roundabout way) means that it's a place (not a person or a thing) where new ideas take root before anyone realizes they exist. Ok, so we won't rehash that, but we want to know where they are. You'll know them when you come across them because these spaces feel different. They move quickly and tolerate chaos. They also reward experimentation, so you can usually sense you're in one when ideas start mutating faster than moderation can keep up.

Edge networks are environments that really thrive on human participation. They're not built around consumption. Instead, they're focused on creation. You'll find them where collaboration still feels organic, like early-stage Discord servers and small Reddit communities. Look at new sub-platforms that are still figuring out what they are. Ultimately, it doesn't really matter which platform or environment you choose to inspect because what defines them isn't the tool or platform. What's really important here is that feedback loop. The faster people can test, respond, test again, and build on top of each other's ideas, the stronger the network becomes.

There are a few reliable ways to tell if you're looking at a true edge environment. The first is low friction, which means it's easy to join, contribute, and remix. The second is signal variation, which means there's no single direction of thought. Ideas split and merge pretty easily. They'll likely reconfigure constantly. The third thing is emergent language. That means inside jokes and new symbols crop up all over the place and shorthand will also start appearing. That's usually the moment a subculture starts coding its identity in plain sight.

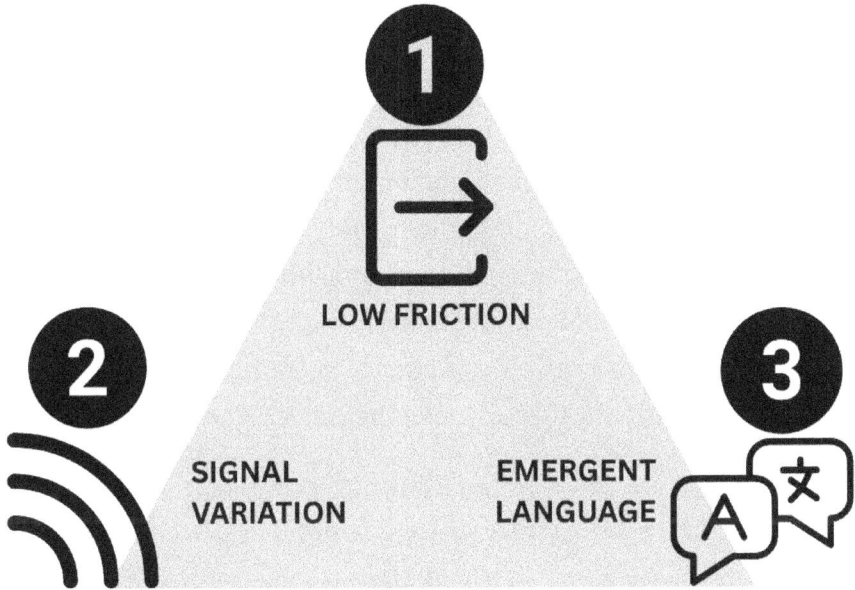

Once those structural signs are visible, then I start reading behavior. Are the conversations emotionally charged? Are ideas being reinterpreted across different contexts? Are multiple creators adopting similar patterns without coordination? Those are the "tells" that something is moving from exploration to early adoption.

Traditional research waits for a report to validate the trend. Edge networks don't. They test in public and evolve in real time. I have a table that's perfect for this coming up in a later chapter.

Memetics & Signal Propagation

Memetics are somewhat linked to emerging language, but when I talk about memetics, I'm not talking about memes in the internet-humor sense of the word. Memetics has more to do with the study of how ideas move. A meme is a packet of meaning that can travel. It can be a phrase, a color, a sound, an image, or a gesture. The number of memes circulating amongst Gen Z-ers at the time of writing this is astronomical and they might make you feel like Father Time. You're not ancient. There's just so much new language. My point is that what makes it powerful isn't the entertainment value, although that's pretty decent. What I look for is the identity signal that's buried inside. When someone repeats a meme, they're revealing what they believe, what they crave, what they agree with, or what they're quietly rejecting. That's why I use memetics to read cultural movement before it becomes measurable.

Memes act like emotional truth detectors because they tend to show up when people feel something before they can explain it. So, in essence, a meme gives shape to what's already there by compressing emotion into something simple and instantly recognizable. When terms like "quiet quitting" or "main character energy" caught fire, they didn't do it because marketers planned it. They spread because people saw themselves in the words. The meme named what they'd already been feeling.

With that said, you shouldn't be counting engagement metrics just yet. I typically start by reading the tone in the meme. I look at how a phrase changes meaning as it travels from one corner of the internet to another. A joke in one group can become a mission statement in another. You know that when irony turns into sincerity, something real is brewing beneath the surface. When a meme jumps between worlds (like from gaming to entrepreneurship or from wellness to productivity), that's when I know it's not just cultural chatter. Now, visuals carry that same charge in the sense that a color scheme or a layout can signal a change in values. When creators started trading glossy perfection for lo-fi authenticity, it was a sign that the general populace or audience had stopped craving aspiration and started craving realness. The visual language of the meme told us that people wanted the truth (warts and all).

So, that's signal propagation through memetics. Now let's talk about memetic spread.

Memetic spread works like a cultural seismograph in that it measures tension and release in real time. A meme doesn't spread because someone tells people to post it. It spreads because it feels true and maybe a little cheeky to boot. That's why I've found that memes predict better than reports. So, to recap before we move on, look at how language mutates across communities. Look for three things.

1. Tone shifts
2. Recurring phrases
3. Emotional undercurrents

I like to pay attention to what gets mocked, because ridicule is usually really good camouflage for desire. These are all just early messages from the future that are hiding inside an internet joke.

Seriously.

Startup Ecosystems & Venture Signals

Next, we have startups. These are the lifeblood of problem-solving in any market. That's why, when I'm trying to understand where the market is heading, I don't start by looking at what big companies are doing. I look at what startups are building. Startups don't come into existence from places of comfort. They're born from a real need. They exist because something isn't working or because traditional players have ignored an opportunity long enough for someone else to grab it. That's why the startup ecosystem is one of the best early indicators of what's about to matter culturally and economically. When enough people are willing to bet their careers and their money on a single idea, that's another signal.

Then, we've got to look at the people funding those startups (if they aren't being bootstrapped) because that will tell you that investors are taking notes.

So, when you read venture capital portfolios properly, you can see that they work like an unofficial schematic of the future. Investors aren't just tossing money around. They or their teams take the time to sniff out what could net them great returns on their investment and that usually means that they're looking for things that will get traction with the right guidance and structuring. They're essentially placing bets on which systems they think will be disrupted or rebuilt. Now, I don't focus on valuations. I'm not a VC. Instead, I focus on patterns of flow. When I see multiple funds backing similar ideas in the same quarter, I know a new category wave is coming through. That's how longevity biotech and AI-driven personalization all surfaced years before they hit mainstream headlines. The market hadn't noticed yet, but the capital already had.

I will say that funding rounds are just one part of a much bigger picture. The next piece of that picture lives in patent filings. From there, we look at something known as accelerator cohorts, as well as research consortiums. Patent velocity for how quickly new ideas are being turned into protectable inventions. Basically, the speed at which innovation is being documented and pushed toward commercialization. Accelerators show us where the most ambitious founders are putting their energy. If I see a couple of startups in the same accelerator building around a similar concept (say, synthetic identity or decentralized infrastructure), that's not a coincidence. What we've got on our hands is conviction and conviction is a better predictor of the future than any press release.

The beauty in this is that startups are also where convergence happens first. Corporates tend to think in categories and founders think in collisions. That's where new industries are born. They'll combine AI with robotics and logistics to rewire how goods move through cities, or they'll merge blockchain and health data into digital medical passports, for example. These combinations create new companies and prototypes for the next version of the economy. Later, big corporations will either buy them or copy them, but by then, the real opportunity will have already moved on for people like you and me.

So, how do you know when that ship has already sailed?

Well, you can preempt it a little. One of the most telling changes I watch for in the startup world is when founders stop building products and start building platforms. That's the tipping point between novelty and scale. It happened with cloud computing. It's happening again right now with AI agent frameworks and synthetic content engines. If you have startups building infrastructure that allows other people to innovate, that's the moment right before mass adoption.

Startups are the best experiments in what the future economy will value. You've got to build a signal radar for it.

Building a Signal Radar

A Signal Radar isn't a piece of software or an antenna I strap to my head before I go out into a lightning storm and pray for inspiration. It might make for a great movie (or even greater meme, depending on how that pans out), but it's not real. A Signal Radar, in my professional experience, is a strategic discipline. It's the system I use to continually do all of that early signal identification. A Signal Radar keeps me operating in the early phase where leverage is highest and competition is lowest.

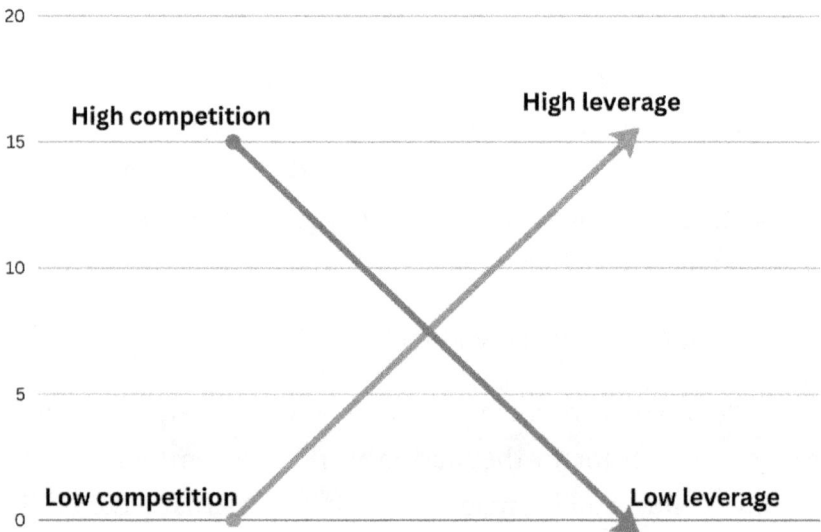

The first step is choosing the right platforms and inputs. I don't rely on mainstream news or trend reports. Those sources only report on what has already happened. Instead, I track fringe platforms, emerging creator communities, patent databases, venture funding announcements, GitHub repositories, TikTok subcultures, and intelligence reports from accelerators. I try to keep a collection of diverse inputs because it lets me see the edges from multiple directions.

Now, to make sense of what I'm seeing, I organize everything into four core categories. Cultural signals show changes in identity, language, and values. Technological signals show me where experimentation is happening. Behavioral signals tell me how people are changing how they spend time, money, attention, and other resources. Economic signals mirror where capital is flowing and where it's quietly exiting.

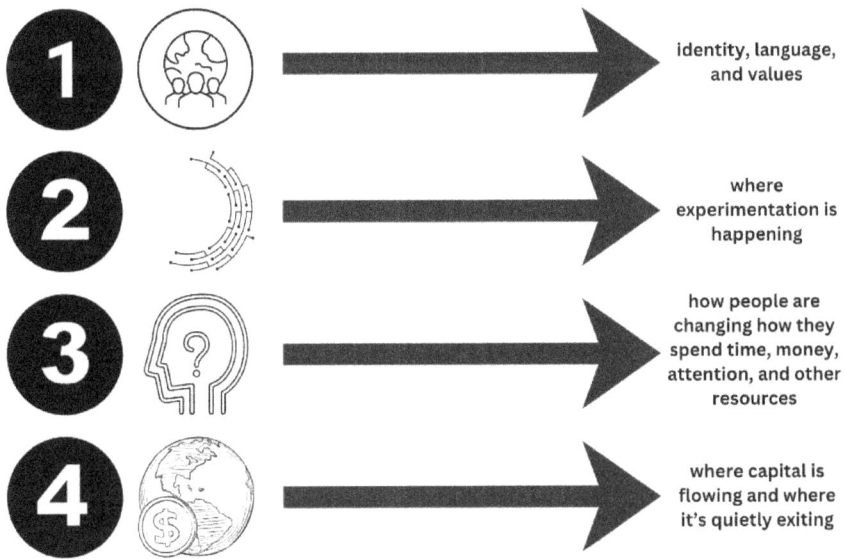

1 → identity, language, and values

2 → where experimentation is happening

3 → how people are changing how they spend time, money, attention, and other resources

4 → where capital is flowing and where it's quietly exiting

This structure gives every observation a place to live. If I didn't have this type of categorization, my signals would just be scattered and I wouldn't be able to garner any meaning from them.

I use the categories for my next phase, which is evaluating my signals using clear criteria. I rank signals based on novelty, which tells me if it represents a new behavior. Velocity shows me whether it is spreading quickly across environments. Emotional resonance tells me if it's tied to identity or aspiration. Scalability is a determinant of whether it can move outside of a niche group and start to influence a larger population.

Just keep in mind that a Signal Radar only works if it becomes a repeatable practice. I log signals frequently depending on the speed of change in a particular category. The more I log the data, the easier it is to see the patterns that come up through repetition. Never forget that the brain isn't designed to remember isolated data points. It's designed to recognize patterns through exposure.

You need to tune into emotional pulses to pick up on those patterns.

Chapter 3 –
The Emotional Pulse of a Signal

Before the viral trend, there's always an emotion, but it's usually subtle. This is a little different from memetics because emotional ignition is not based on (or extracted from) entertainment or curiosity. It carries a different quality and it feels charged. Emotion always shows up before logic. People feel the future before they understand it and excitement is one of the clearest ignition points.

When I'm tracking cultural ignition, I look for the moment when people stop being just curious and start feeling something. You can see it in their language. Rebellion works the same way. When people start pushing back against an old system or aesthetic, it's a sign of readiness for something new. I'll get to what that new thing might be and how you figure the finer details out, but what you should know now is that they're emotionally done with what exists and are open to what's next. (Yes, even if they haven't seen it yet.)

Frustration is another early tell because it means a norm has hit its ceiling. When enough people are irritated by a current process or product, they start scanning their environment for alternatives. Sometimes, that's consciously and sometimes it's not. Then there's belonging, which I find is the most reliable predictor of all. When people find something that makes them feel seen or connected to a new identity, adoption tends to move fast. They're binding their sense of self to a new idea and that's worth more than just engaging with it.

I can't stress this enough. Emotion always fires off first before adoption.

Emotional Feedback Systems

The future isn't built on isolated feelings. We build on emotional feedback systems. These systems take shape when people see others reacting with a specific emotional charge and when that charge bounces back, it reinforces itself before it creates this type of shared psychological field. Now we can see that this is when a loose signal starts evolving into a cultural force.

As I said, belonging is one of the strongest amplifiers in that process (Brewer, 1991). When people feel that a new idea is going to give them access to a rising identity group, they join it. It's not usually a conscious decision and that's why we refer to it as signaling. People want to feel like they're part of what's new and not like they're clinging to what's fading away. It's part of human instinct to want to belong and it turns all of these scattered emotional sparks into collective energy.

From there, validation amplifies it.

When someone sees their excitement or frustration reflected back at them, it confirms that they weren't the only one feeling it. Suddenly, that emotion turns into a form of evidence that something real is happening. That's how emotional reinforcement begins. The more people see other people feeling the same thing, the louder they get and the more they share. That's how the narrative spreads even faster.

Next comes mimicry. Humans are wired to copy emotional states (Prochazkova, 2017). One person's enthusiasm or outrage becomes contagious. Other people pick it up subconsciously and the emotion starts moving on its own. That's what pushes signals out of small communities and into the broader culture. Once that bit of resonance becomes collective, it stops being about individual reaction and starts becoming a channel for behavior.

At that point, the emotional system takes over. People aren't responding to the original trigger anymore because they're actually responding to the energy circulating around it. That's how cultural momentum forms and these are the feedback loops of shared emotion that tell me we're at the start of psychological adoption. The cascade will look something like this.

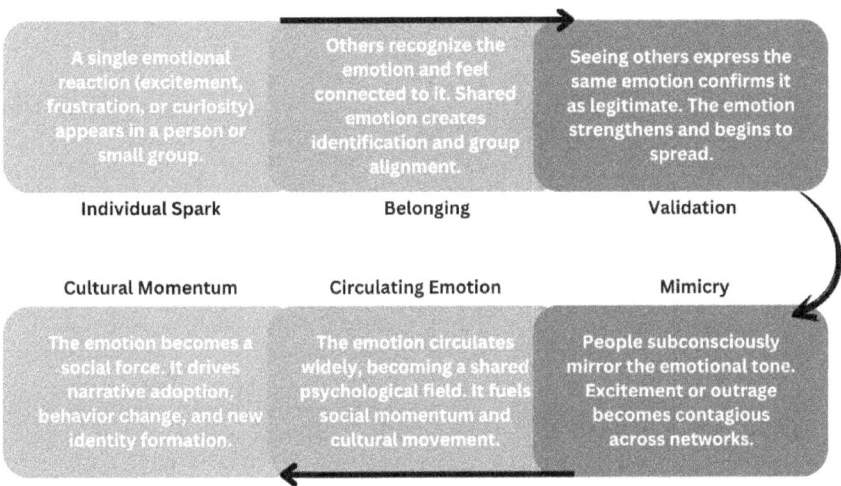

Individual Spark	Belonging	Validation
A single emotional reaction (excitement, frustration, or curiosity) appears in a person or small group.	Others recognize the emotion and feel connected to it. Shared emotion creates identification and group alignment.	Seeing others express the same emotion confirms it as legitimate. The emotion strengthens and begins to spread.

Cultural Momentum	Circulating Emotion	Mimicry
The emotion becomes a social force. It drives narrative adoption, behavior change, and new identity formation.	The emotion circulates widely, becoming a shared psychological field. It fuels social momentum and cultural movement.	People subconsciously mirror the emotional tone. Excitement or outrage becomes contagious across networks.

When Signals Become Self-Expression

The next thing I want to do when I'm tracking these emerging changes is watch for self-expression. I know that many leaders reading this are probably wondering why I didn't say adoption and it's for good reason. Adoption is shallow and self-expression is deep. What I mean is that a person might try a new app or product because it seems like it's useful or interesting, but they only commit to it if it becomes part of their identity. This is the moment when a signal moves from what we'll call novelty to inevitability. People don't adopt trends. They adopt identities *through* trends. So, what looks like consumer behavior on the surface is actually just self-definition being played out in public.

In essence, every future movement begins when people use signals to communicate who they're becoming. A signal can be any of the following.

1. Visual aesthetic
2. Phrase or vocabulary
3. Style of interaction
4. Lifestyle choice
5. Digital presence
6. Purchasing behavior
7. Technology adoption
8. Symbolic affiliation
9. Spatial alignment
10. Creative expression
11. Moral stance
12. Financial behavior

These aren't superficial because when someone changes the way they speak online or chooses a specific visual language, they're signaling alignment with a future narrative.

Now, utility might be what gets people curious at first because a technology that saves time or makes something easier will always draw people's attention. However, usefulness alone doesn't build movements. Remember that when an idea reflects who someone wants to become (even if they can't name that version of themselves yet), it stays with them. People don't come back to products or communities because they're efficient. Efficiency is great, but most of us don't really buy into it unless there's something underneath. So, people come back because being part of that collective feels like a step toward the person they believe they're supposed to be.

This is where marketers often miss the turn. They measure clicks and signups, but they miss when a product or trend stops being functional and starts being symbolic. When people begin using something as a way to express who they are, persuasion doesn't really matter anymore. Features don't count for much either. At that point, a tribe forms around a shared belief in a story about the future.

When I see people defending a trend because it *means* something to them, I know the signal has passed the threshold of utility. That's when it becomes unstoppable. Features can be copied and technologies can be replaced but the moment a signal becomes a part of how people see themselves, it no longer competes in a category because there is no competition.

Invisible Operating System

Values are what I like to refer to as the invisible operating system behind human behavior and that's because they sit underneath emotion and quietly run the code that decides which feelings take root. It's important to understand what a value really is. It isn't a preference or some pie-in-the-sky ideal that a person just decides on having one random Monday morning. It's part of a greater belief system that's been shaped over time and tells its user (or believer) what should matter to them. Values reshape how people decide to spend and how they define themselves, which means, eventually, it redirects the path of entire industries.

I'll give you a couple of examples.

Value	Sector or Business
Sovereignty	decentralized finance, personal AI tools, privacy-first tech, freelance ecosystems, off-grid living products
Longevity	biotech therapies, supplements, wearables, functional fitness, regenerative medicine, precision nutrition
Anonymity	encrypted platforms, data protection software, burner identities, stealth luxury, digital minimalism
Authenticity	handmade goods, transparent supply chains, raw or unfiltered content, community-led brands
Freedom	remote work systems, van-life gear, travel-as-lifestyle services, minimalism, digital nomad infrastructure
Sustainability	circular fashion, electric transport, carbon tracking, renewable home systems, zero-waste design

Table 2: Values and Their Sectors

Values are psychological priorities that dictate a person's buying patterns and that surface through emotion before they show up in data. When people talk about sovereignty, they're not just talking about making money online or working from anywhere. They're saying that control over their own lives now feels non-negotiable. When longevity becomes a value, people start optimizing for health like it's a moral duty and not just a lifestyle choice. Desire for anonymity is a sign that a big enough subset of people is fatigued from overexposure of some sort.

These values tell me where the economy is going long before the market reports catch up. If sovereignty is rising, I expect to see growth in decentralized tools and systems that cut reliance on institutions. If longevity is rising, the expansion will happen around biotech and personalized health or wellness ecosystems. First comes the emotion, then come the values, and then the markets just follow their lead. Tracking them is early-stage economic forecasting.

It's all in the push-pull of emotional tension. When an outdated value is being challenged, people begin to express frustration or fatigue. When a new value is coming up in its place, I see emotional excitement or aspiration. At some point, those two lines intersect and there's tension between what is and what should be. You can usually calculate the trajectory of a value and whether it's either ascending or declining pretty well in that spot.

Values function like gravity in that they pull everything toward them, including technology, business models, cultural narratives, and brand relevance. Once a value gets set in motion, the future is already in motion with it.

Desire Mapping

So, we've got emotional feedback and values as an operating system. Now, we need to circle all the way back to desire. Every future shift I have ever tracked begins with desire. Before a product is created and well before a narrative takes its shape or becomes visible, there's a human need that's on the hunt for a new path of expression. Remember, emotional ties aren't logical, which means humans don't move toward technology or cultural change for logical reasons. They move because something inside them wants more of something or wants freedom from something. That's why desire is the next reliable predictor of future momentum.

There are five core human desires that drive long-term market behavior.

1	2	3	4	5
Status	Control	Belonging	Novelty	Safety

Every emerging signal eventually maps back to one or more of these roots. So, when I'm in the process of identifying a signal, I immediately ask which desire it's serving. I don't care if it's a product, a meme, a movement, or a financial behavior. It won't scale if there's no desire attached to it. Simply put, the signal will stall if the desire is weak.

Now, let's talk about each one.

Status drives people to adopt new things that signal leadership or importance. This is why luxury evolves faster than any other category. Control drives adoption in areas like financial independence and self-sovereign systems. You'll find that central intermediaries become obsolete when control rises as a priority. Belonging drives subcultures and community-led brands and when belonging becomes scarce in the physical world, it explodes in digital spaces. Novelty pulls people toward the future when they feel stuck in the present. It's the root desire behind invention and experimentation. You'll find a lot of it in youth culture. Safety drives everything from cybersecurity to biohacking, so it's not only based on physical safety, but emotional and reputational safety, too.

In essence, there's less of a focus on what's interesting and more of a focus on which human desire we're unlocking. A trend with no desire mapping won't last. Desire follows that emotional gravity, which is why people will pay for status before they pay for utility. They will choose control over convenience and will sacrifice a reasonable amount of profit for belonging.

The point is that desire mapping is how I determine if a weak signal is just an anomaly or a future inevitability.

Perception Extension with AI

How do I pull all of this together in a world that is becoming exceedingly fast and demanding? The answer is AI, which I know triggers many of us in these fields, but it's a tool. So, when it comes to seeing the future, I don't look at AI as replacing my ability to perceive. It expands it in the same way that wearing glasses would let me see better. My brain is wired to detect things like emotion, narrative, desire, and meaning, but it operates linearly. I can pretty confidently say that almost all of us (if not all of us) have a brain that works that way. AI operates exponentially. It sees everything all at once, which means it can process millions of conversations and evolve models in real time in order to detect a subtle change in language. Plus, it can do that before anyone even notices that something has changed. What AI gives me is enhanced perception. It shows me what's rising beneath awareness. Of course, you have to know what you don't know. You have to know how to prompt it and exactly what you're looking for in the first place.

The real power of AI in prediction isn't just speed because it doesn't just produce more data. It takes that data and exposes patterns that are too subtle or too deeply buried for the human brain to catch on its own.

I use AI to map emotional trajectories across digital conversations. It detects the exact moment curiosity becomes a movement. I might have a keen eye for spotting how skepticism gets transformed into rebellion, but I'll never, for all my years of experience, be able to do it as fast as a machine. There are a handful of people who calculate mammoth equations in seconds, but they're still not faster than the average desktop calculator that's collecting dust in your dad's home office. Basically, I'm reiterating for the third time that this is a tool. Those emotional transitions are the earliest markers of mass behavior change and it's so important to catch them early. AI helps me see that while it's still barely a murmur and long before it's visible to the mainstream.

AI also acts like a mirror for collective desire. It doesn't tell me what people want, but it's quite good at reflecting what they're already signaling by sifting through language patterns that point to deeper needs. It picks up rising interest in things like longevity, anonymity, digital sovereignty, emotional liberation, and self-optimization.

Still, AI can't tell me *why* those signals matter. That's still the human part. We have to find meaning in what it brings to our attention and meaning comes from interpretation. It's not a computational process just yet. So, to sum up, the model can detect emotional movement, but I get to decide what it means and whether it's worth acting on. I like to think of it as machine detection meeting human discernment.

There's a simple ten-step process that I typically use to feed the STEEP framework that I'll be introducing you to in the next chapter. So, let's move on to how I actually turn data into meaning using AI and other more conventional tools.

Part II: **U**nderstand Meaning

(Encode)

Chapter 4 –
Turning Data into Meaning

So, this chapter is going to follow along the previous one pretty closely. Rome wasn't built in a day, after all. The reason for it is that I need to break this down into steps that are as easy to implement as possible. I'm essentially giving you a four-year marketing degree in a couple thousand words, so I have to be deliberate here. The difference here is that instead of finding signals and their meaning, we're going to focus on understanding that meaning.

I read somewhere that we're drowning in information but starving for insight, and I feel that's true. Every platform generates endless streams of data, but most organizations are still making decisions based on the noise that we discussed earlier. Raw data is practically meaningless, which simply means that information alone doesn't drive transformation. What creates a competitive advantage is the ability to extract meaning from whatever data you're getting. Ultimately, meaning is what tells you why something matters and what it is leading to next. That's the real power of future readiness.

Raw data is just unprocessed input. It's everything from keyword signals and search behavior to consumer spending and social engagement metrics.

Information, on the other hand, is organized data. It tells you what's happening on the surface.

Meaning explains why it's happening and what it reveals about underlying human behavior.

Think of it like a sandwich. If you slap some bread and a few condiments down on the table, you can't just eat them as they stand. You have to assemble them to get something useful out of them.

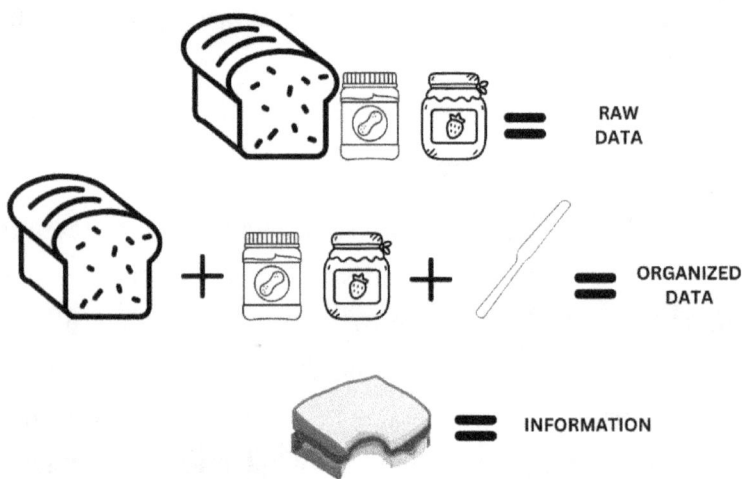

This is why when I'm scanning signals, I'm not asking what people are doing. I'm asking what their behavior means in terms of identity shifts, psychological drivers, or emerging needs. Meaning is what lets me see the future before it arrives.

The issue is that most leaders believe they have a data problem. They think that if they had more reports or better analytics, they'd make better decisions. In reality, the problem is interpretation. Two leaders can look at the same data and arrive at opposite conclusions depending on how they frame the signal. The ones who win are the ones who know how to translate information into actionable meaning. This is why insight is more valuable than intelligence. Intelligence tells you what exists. Insight tells you what's coming.

Meaning comes from interpretation. It's the mental process of pulling threads together across time, context, behavior, and psychology. When you interpret a weak signal correctly, you're both seeing what's happening in the moment and seeing what that moment is going to set in motion. That's the entire difference. Data can only point to a direction once the change has already found a spot to settle into. Meaning shows the direction while that change is still forming.

Leaders who understand the jump from information to insight stop acting like reporters and start acting like pattern readers. When you can look at behavior and ask what emotion is powering it or look at new platforms and ask what desire is being expressed, you're on the right path. For that, you've got to go back to the previous chapters and look at how to treat early adoption as a clue about identity evolution and not treat it as a trend chart. When you flip your mindset like that, you'll be able to create actual predictive ability.

Use STEEP Framework for Meaning-Making

Of course, that's all easier said than done. I can tell you to find meaning all day long and I'll probably come off like some spiritual guru that's lecturing you on how to find your purpose. It's going to sound like another language to you if you've never really done it before. Frameworks are what you need to draw that meaning out.

The one thing that I won't ever do when I come across an emerging signal is accept it at face value. I run it through a structured lens to figure out what's actually underneath. The STEEP framework is one of the clearest and most reliable ways to do that. It forces me to stop obsessing over what the signal looks like on the surface and dig into the forces shaping it behind the scenes.

STEEP stands for Social, Technological, Economic, Environmental, and Political. These are the five pressure systems that quietly influence every trend, preference, and behavior pattern we see. When I use it, I'm categorizing data, but I'm also decoding what the signal is trying to tell me about the direction the world is leaning toward.

S SOCIAL T TECH E ECONOMY E ENVIRONMENT P POLITICAL

The Social lens is the first place I look because culture always moves before markets do (Aguilar, 1967). This lens helps me see whether a signal is tied to changing values, new identity needs, demographic movement, or changes in how people connect. Social change is almost always the root cause. For example, if I see a surge in platforms built around anonymity, I don't write it off as a passing preference. I read it as a social response to identity exhaustion, hyper-visibility, safety concerns, or the need for psychological breathing room. Social signals are early indicators of cultural reorientation, and when they show up, the rest of the system usually follows.

The Technological lens shows me how infrastructure and capability are opening new doors for behavior. Technology doesn't just hand people new tools. It's actually the thing that rewires what they think is possible. It changes their expectations as well as their threshold for convenience and control. So when I notice multiple startups building around autonomous logistics, I don't see "delivery innovation." I see a deeper technological driver. For me, I might say that it's the expectation that frictionless speed is becoming the new baseline. Technology tells you what people can do and what they'll soon expect as normal.

The Economic lens is where I measure the pressure points that shape adoption. These are affordability, access, resource constraints, and the flow of capital. If a behavior spikes during a moment of inflation, stagnation, or uncertainty, I know it's a coping mechanism. If you don't know this, coping mechanisms often mature into permanent behaviors once they're reinforced by economic necessity. When the economic environment evolves, people change their choices and their priorities. Economic signals matter because they show you what behaviors will stick even after the crisis passes.

The Environmental lens covers the entire physical reality a signal has to move through, including resource availability, ecological pressure, geographic limitations, and the basic constraints of the real world. When supply chains collapse or materials become scarce, behavior moves because it has to. Localized production[1] and regenerative systems[2] rise because environmental pressure forces innovation faster than cultural enthusiasm. The same goes for circular economies[3]. The point is that environmental signals accelerate adoption by sheer necessity and necessity is one of the most powerful engines of change we have.

Lastly, the Political lens tracks the forces that most people underestimate until it's too late. These are regulations, policy shifts, global tension, and changes in institutional power. A signal can gain momentum or collapse overnight depending on how political structures react to it. Sometimes a trend takes off because the public is obsessed with it and legislation clears the path. The opposite is also true because other times, a promising movement stalls because policy walls go up around it. Politics is one of the strongest catalysts for mass acceleration in sectors like energy,

---

[1] The creation of goods within the same geographic area where they are consumed, reducing reliance on long supply chains and increasing resilience.

[2] Processes that restore, renew, or revitalize their own energy and materials, producing more value and resources than they consume.

[3] Economic systems that keep materials in use for as long as possible through recycling, reuse, and design for longevity rather than disposal.

biotech, health, and finance. If the political environment moves, the market moves with it.

So, when I apply STEEP, I'm turning a single data point into a full narrative about where the future is leaning. Telemedicine is a perfect example. On the surface, it looked like a convenience play, but its rise was driven by a social need for broader access, technological readiness through broadband and video platforms, economic pressure to reduce healthcare costs, environmental constraints during a global pandemic, and political deregulation that finally allowed remote care at scale.

I'll give you another example with Ozempic and GLP-1 drugs.

1. SOCIAL

People were exhausted by diet culture and frustrated with traditional weight-loss systems that required time, discipline, and emotional labor. There was also a rising belief that health optimization should be accessible, fast, and science-driven.

2. TECHNOLOGICAL

GLP-1 medications improved dramatically. AI-accelerated drug discovery sped up clinical refinement. Those same telemedicine platforms that kicked off during the pandemic allowed prescriptions without in-person visits. At-home monitoring and biomarker tracking made treatment easier. Technology removed friction at every step. (Wang JY, 2023)

3. ECONOMIC

The weight-loss industry was a $70B market that was desperate for disruption. Consumers were willing to pay out of pocket. Investors poured billions into biotech firms building GLP-1 competitors. Employers and insurers later stepped in because healthier employees meant lower long-term costs.

4. ENVIRONMENTAL

The pandemic created massive lifestyle disruption (Giuntella O, 2021). Stress-related weight gain, limited movement, and health fears amplified demand. Healthcare systems were overloaded, which meant that fast pharmaceutical solutions became more appealing than long-term programs.

5. POLITICAL

Governments loosened telemedicine laws during COVID. Insurance frameworks began adapting to include GLP-1 coverage. Regulatory fast-tracking of obesity treatments signaled political acceptance. Policy removed bottlenecks and legitimized the entire category.

There's a clear path that this all followed and whether you want to hop on the conspiracy theory bandwagon of why COVID came to be or not, the fact is that it disrupted a lot. There's good and bad to that. At the end of the day, what matters is that people look for implements to impact their lives positively or to do a job for them.

"Jobs-to-Be-Done" for Human Motivation

So, if people latched onto GLP-1s because they did a job for them (replaced having to go to the gym or plan a diet), then it goes without saying that people don't buy products. They hire them to perform a job in their life and that job is always tied to a deeper desire or problem that they're trying to solve. If I only look at what someone's using, I miss the bigger truth, which is hidden in why they're using it and what job it's performing emotionally or functionally.

A job isn't a task. It's an outcome someone is trying to feel or resolve. Nobody "hires" a fitness app because they want charts and data. They hire it because they want to feel in control of their health or because tracking their progress reinforces a sense of discipline. The same logic applies to luxury brands. People don't buy a designer bag because they need something to carry their keys in. They buy it because it gives them status or belonging inside a social hierarchy. Essentially, everything does a job, so you have to understand what job it's really doing.

Jobs-to-Be-Done lets me interpret signals through human motivation instead of surface-level behavior (Christensen, 1997). This lens also helps me separate novelty from genuine demand when I connect it to the insight from STEEP. If a signal satisfies an entertainment job, it may spike and fade, but if it satisfies a job tied to identity, safety, status, or personal transformation, it carries strategic weight.

So when I evaluate a weak signal, I ask one question. What job is this product, behavior, or technology being hired to do? If the job is emotionally charged and underserved, adoption will move.

Pattern Framing & Sensemaking

You still need to frame the patterns to make sense out of all of that organized data. Without it, you might just end up with the peanut butter and jelly outside of the bread instead of between two slices.

Pattern framing is the process I use to connect signals that look unrelated on the surface but are actually expressions of the same emerging force. The brain is built for pattern recognition. It's evolutionary because recognizing patterns allowed us to anticipate danger and survive. It's why we can stare up at clouds and see bunnies or see faces in the patterns on our hotel drapes. Our brains were wired to see the pattern of a predator's face even if it was hiding in the brush (Rienzi, 2024). That wiring is called the pareidolia phenomenon and doesn't just live and die with safety. It's also what helps us see other patterns.

I'll stop talking in circles now.

In foresight, pattern framing works the same way. It helps me move past categories like "technology trend" or "consumer trend" and focus on the deeper momentum shaping them both. So when I notice off-grid living, biohacking, decentralized finance, and homeschooling rising at the same time, I don't treat them as four separate movements. When I frame the pattern, I see one force. In this case, it might be a growing desire for sovereignty and control over personal systems.

I start by watching for repeating emotional or behavioral themes. Is there more language around autonomy, safety, identity, or self-governance? Are people choosing alternatives outside institutional control? Once those patterns show up, I cluster them into proto-trends, which is just a technical term for early formations that haven't fully matured but clearly point toward a new archetype emerging in society. These archetypes become the psychological stories people start aligning with before the behavior goes mainstream. When I can see the archetype forming, I can see the direction of the future long before the market catches up.

That's why I use all of these frameworks. I can't stress their importance enough.

When I'm mapping emerging patterns, I sort signals by their emotional drivers, their speed of adoption, and the level of disruption they pose to existing systems. If several signals are responding to the same emotional need and their velocity keeps increasing, that's the early shape of a proto-trend with real future impact.

The real strategic edge comes from telling the difference between isolated events and emerging archetypes. An isolated event can make headlines, but without a deeper emotional or behavioral pattern beneath it, it dies quickly. An emerging archetype, on the other hand, shows up everywhere as variations of the same psychological story.

In short, pattern framing isn't prediction. It's perception at scale. It turns scattered signals into coherent direction and gives me a privileged view of what the future is preparing to reward next.

A company that knows how to define meaning doesn't sit around waiting for the market to validate its choices. It sets the tone and shapes the conversation. Then, it frames the narrative that everyone else eventually repeats. That's how categories get created and why some brands always feel a step ahead. It's not necessarily because they have supernatural instinct. It might look that way, but everything we once thought was supernatural or magical almost always turned out to be a system that we just couldn't see. This system gives companies the edge to interpret signals earlier and position themselves as the ones naming the shift. They're authoring trends instead of reacting to them. So, no, there isn't someone in the back with a cauldron and a crystal ball. We can put our pitchforks and torches down.

Meaning-making takes two things.

1. Perceptual awareness

2. Narrative clarity

When I'm training this in myself, I push past surface-level observation and ask deeper questions every time I encounter a new signal. These questions force me out of description mode and into interpretation mode, which is where strategy lives.

Companies that operate only from data end up optimizing for efficiency. They get faster, but they don't get ahead. They move with the market instead of moving the market. Companies that look like they hold the crystal ball build their offerings around the story they want consumers to believe in. That's the difference between becoming a category leader and becoming another product in a crowded aisle.

To build meaning-making as a personal discipline, I set aside intentional time each week to study not just what I'm noticing but how I'm interpreting it. I log signals and then I make myself articulate their deeper implications. I connect dots across unrelated industries. I map the narratives forming under the surface and track which identities those narratives are shaping.

With enough practice, this trains your brain to stop collecting information and start extracting insight, which is the real currency of foresight. This has to be done because in a world where AI can surface unlimited information, the human advantage is always going to be in finding the meaning.

Chapter 5 –
Insight Architecture

The next step in understanding meaning is taking what you've collated using STEEP and creating insight architecture. It sounds like something a biz-dev consultant would pitch in a mahogany-covered boardroom, but it's actually simple. Most people collect signals the way they collect browser tabs. They'll play around with frameworks like STEEP, but they'll still fail to see the real behavior behind any of it. That's where the structure pays off because it gives your brain a clean path from raw observation to real insight.

Real options thinking supports this because you're not looking for one final answer. Instead, you're giving yourself a set of possible moves that stay open while the pattern forms. It works like checking a handful of routes on a map before you choose the one that gets you to the destination the fastest. So, if STEEP helps you deduce whether one or more emotional changes are turning into something substantial, insight architecture will be the next level up from that. What I mean by that is that it will narrow down your choices.

For this, we need to unpack framing a little more because it will let you group signals by the emotional drivers behind them. You also group them by shifts in identity because people will show you who they're becoming long before they say it out loud. Then, you'll need to add in or account for the new tech capabilities that are changing what people believe is possible. You'll also add in the behavioral changes that surface when enough people get tired of the old thing and start leaning toward the new thing. You have to keep your eye out for the moment when these frames or clusters repeat themselves in different corners of culture.

What you're looking for is the moment of convergence, which you can plot on a meaning map.

Mapping Meaning

When you've run your signals through a framework and you've narrowed them down, you'll want to create a Meaning Map as part of your insight architecture. This map helps you turn early signals into something you can actually work with. You do this by laying the signals out in a way that shows how they relate to each other instead of treating them like scattered dots on a page. The real insight (which is the step after your PB&J is made) only shows up when you start working through the connections. However, those connections can only become visible on your Meaning Map.

You build a Meaning Map through a simple step-by-step flow.

The first step is gathering the signals that seem important.

The next step is spreading them out so you can see them as separate pieces.

Then you start grouping them by the qualities they share.

Emotional energy is the first clue you look for. You already know what those look like from all of the previous chapters (head back if you need a refresher). You place signals with high emotional energy on one area of the map. The low-energy signals sit in another area that you've chosen.

The next step is adding urgency. Some signals move fast and jump across communities with real speed, but not all of them are like that. Urgency helps you tell which ones might play a role in shaping near-term behavior and which ones might be a reflection of slow cultural change. You place fast-rising signals in the upper half of your map and you place slower ones lower. You're giving them positions so you can read the direction of movement.

After urgency, you add market relevance. This tells you whether something is influencing what people value, how they spend, how they choose products, or how they decide on experiences. A signal that has strong market relevance should sit close to the center of the map because it touches real economic behavior. A signal with little relevance should sit further away from the center. Once you set these positions, you can see which clusters have emotional charge, urgency, and economic pull. These clusters are the ones worth studying because they point to the direction in which future demand will form.

This visual layout breaks your personal bias. No matter how trained you are, every person has assumptions that distort their judgment. When everything sits in your head, those assumptions win. When everything sits on the map, the truth wins. You can see which signals show up across categories instead of relying on what feels familiar to you. Trust me, this is going to save you from overlooking a weak signal too early or treating a loud trend as a long-term force.

The next step is deciding whether a movement is surface-level or structural. Remember, surface-level signals are trends that rise fast and fall fast. They change language instead of identity. They create conversations instead of new behavior. Structural signals form deeper roots because they're linked to a human need or a rising value. You'll know it's a structural trend when it shows up in several clusters on your map. When a signal keeps touching emotion, urgency, and economic behavior, it's structural.

That's when you pay attention.

Finally, you'll layer pattern language into the map. Pattern language gives shape and names to the clusters that you've identified. It also gives you a way to talk about these forces without slipping into complicated jargon. Ideally, you should name each cluster based on the driver behind it. You can start with the desire or value at the root and attach the behavior that's linked to that desire if that helps. From there, it'll be easier to note down the domain where the signal is taking shape. A cluster tied to identity protection inside creator spaces might become Belonging Pseudonymous Presence Creator Economy. When you say it out loud, you understand what you're looking at. You see the emotional driver as well as the behavior and the domain. That means you'll likely also understand the direction at that point.

Driver anchored naming gives you a plain-English system that you can apply across industries. You can use the same label in marketing, product, leadership, or investment conversations because it tells you what the signal means without forcing anyone to translate it. Keeping your vocabulary consistent means that it becomes a shared tool.

Once you add these names to the Meaning Map, you get a clean language system that flows through your strategy. You might think it's fluff, but it helps teams make decisions faster because everyone understands the same labels. The map becomes reusable for planning documents, leadership decks, workshops, and courses. You also build a library of named drivers that you can reference as those signals evolve. This gives you a long-term structure for reading the world instead of reacting to whatever shows up online this week.

Minimum Viable Bets

Once you've got your Meaning Map, you need to move by placing a minimum viable bet. This is the smallest move you can make that still teaches you something real about the market. Most teams wait for the perfect plan and end up in a tailspin of polishing or proofreading documents. You can't keep adding layers until the idea feels safe because safety is a feeling and intuition is the only hunch we need to be going on (and only in the very early stages). If you get stuck in this type of failure to launch, you'll only get to the market when you're fighting to find an angle or being actively priced out by those who tapped in sooner.

A minimum viable bet keeps you from falling into that trap because it pushes you to take action before the idea hardens. You learn by doing, not by planning. (Well, maybe with a little bit of planning.)

So, let's get you on to your first minimum viable bet. You start by identifying the smallest move that can confirm or challenge your strategic truth. The size doesn't matter. A good minimum viable bet will show you whether people care and understand the value. You'll be able to tell fairly quickly if the behavior you predicted can actually show up in the wild. You avoid big infrastructure and heavy builds, too, which is always a plus. The point is to aim for the fastest path to a real reaction. If the market leans in when you make that bet, you know the strategic truth has legs. If the market leans out, you can adjust before you waste any more of your time.

How do you accomplish, though? Well, you have several tools for this.

A test launch is the simplest. You put a small version of the idea in front of a real audience and measure the pull. You shouldn't be looking for mass adoption at this stage. Try to find the signs of early traction. A micro experiment works the same way because you get to release something small and watch how people respond to it. It might be a landing page or a new feature. That little pivot in positioning isn't to impress anyone. It's a learning opportunity.

Speaking of which, positioning sparks are also important. A spark is a tiny message that introduces the idea without a full rollout. You use it to test whether the emotional driver behind the strategic truth connects with people. If the message spreads on its own, you know the insight is strong. Apple did this in 2019 with their "What happens on your iPhone, stays on your iPhone" billboard across the street from the Consumer Electronics Show at the Las Vegas Convention Center (Haselton, 2019). It was a pretty bold move at a time when people were getting exceedingly more alert about how big tech was storing and using their data. Just the year before, Mark Zuckerberg had been brought before the US Congress for the Cambridge Analytica data scandal (Feiner, 2019). So, Apple showing up across the street from one of the biggest conventions for notoriously data hungry companies and organizations was calculated. They were using some cheeky wit to test the market with a positioning spark.

Prototype experiences are another option. You build a small sample of the future experience and let people try it. You watch where they get excited and where they hesitate. Before Airbnb committed to building Experiences, they tested a small prototype where a few hosts in one city offered mini tours or local activities.

The point of a minimum viable bet is speed of learning. Speed beats scale because scale only helps you when you already know you're right. Speed helps you when you're still figuring things out. A small test gives you answers while the idea is still flexible. It has to come before a big launch because a big launch locks you into a commitment you might not be ready for.

When you're at this level in the process chain, you should be judging your success by what you learn and not by how many people show up to use your product or feature. Even a small group can tell you everything you need to know if the test is clean because you're looking for proof of desire and value through behavior.

The market always tells the truth and a minimum viable bet helps you hear it faster.

Narrative Prototyping

I spoke about prototyping a couple of paragraphs ago and we're going to focus on that but from a narrative perspective. I'm all for narrative prototyping for the simple fact that it's one of the easiest ways to test a strategic truth because you don't need a product or a full plan. You only need a story that shows you the world that your insight describes. A story moves faster than a feature set because people understand narrative on instinct, so when you share a future story, you place your audience inside the world your insight points to. You're putting your feelers out there and, in turn, you're letting them feel the shift before anything is produced.

Try this. Write a simple story about the future state that your insight supports. Create a narrative that explains what life looks like when the strategic truth plays out. Try to focus on the tension driving the change and highlight the behavior that rises from that tension. The next part is very important because (you guessed it) it builds even further on top of what you've already learned in previous chapters. In this step, you need to show how the person in the story wins or avoids pain as a result of your new solution's existence. This step teaches you how to frame the story so that the public understands the value of your solution without you forcing it on them. When you show them how the person in the story wins or avoids pain, you teach the audience how to interpret the idea. You guide the reaction instead of leaving it up to chance and you might just show them a problem that they didn't even realize (consciously) that they had.

After you've shaped the story, you release it into the world. You can share it with customers, partners, teams, or niche communities that are tied directly to your space. Then, you watch what the story triggers in terms of reactions.

Features only matter when they support the desire that you'll see through their reactions. When you use narrative to express the future that your insight points toward, you'll see whether the emotion at the center connects. If it does, you have the validation you need to move forward. In the short term, this is going to help you catch blind spots early because people will probably question parts of the story. Don't freak out! They're giving you the edges of the insight that might need some shaping. They're telling you what feels unclear to them and how they see the world changing in ways you haven't considered. You're getting to strengthen the insight without being weighed down by production yet.

Ultimately, this gives you a low cost and high impact path to clarity.

Operationalizing the Insight

You've gone through frameworks and architecture. You've even mapped the meaning and made a minimum viable bet (maybe with a narrative prototype), but you're only just starting to roll up your sleeves. Once you have a strategic truth, you need to turn it into a move that the market can feel. Operationalizing an insight is the opposite of reacting to a trend because when you do it, you're not following a direction. You're letting the insight guide your next step before anyone else sees what's forming so that you can create the direction.

When the prototyping confirms your educated hunch and you've refined the edges, you'll move on to sequencing. You don't jump straight from idea to large scale launch. You move through a simple flow that started with the prototype and then moves into early traction. You might expand the test audience or open a small beta to roll out a light version of the experience.

Once you have enough feedback, you get to decide whether the insight becomes a permanent strategic direction. This decision comes from real time reactions. Do not, and I mean that as definitively as I can possibly say those two words, base anything on internal excitement. People tell you the truth through what they click, share, buy, question, or ignore. You study the patterns. If the insight continues to produce consistent behavior, you scale. If the response becomes soft or confused, you pivot. Now's not the time to be stubborn.

You need to stay calm and collected because you need to exhibit leadership continuity. I can't tell you how many insights die because teams fracture around interpretation. You avoid that by creating a simple narrative that explains the insight and how it guides action. Then, you share that narrative across teams so no one builds in isolation. Everyone moves toward the same point on the horizon.

You also keep communication open as the insight evolves (and use the language you built in the mapping stage). Thank me later because the work will stay coherent instead of breaking into pieces. From there, you transform.

Part III: **T**ransform Assumptions

(Unlearn)

Chapter 6 –
The Neuroscience of Unlearning

You've found the signals and drawn the meaning out of them, which means that you now have the tall task of transforming your assumptions about all of the information in front of you. Thankfully for you (and me), the human brain has something called neuroplasticity. That's the brain's ability to change its own structure. It's the reason you can learn new skills, form new habits, drop old patterns, and respond to new environments without getting stuck in the past. Up until recently, talks around plasticity had it pegged as a soft science idea. It's not. It's a physical property of your nervous system and it works to your benefit from a marketing standpoint. When the world changes faster than your routines can keep up, plasticity helps you adapt to said changes. You would probably repeat the same response long after reality has moved on if you didn't have it.

I've given you a similar analogy before, but I doubt you're going to trek all the way back to Part I to find it, so I'll bring it back home now. You can think of plasticity like the way a trail forms through a forest. Every time you walk that path, the ground gets a little clearer. If you walk it every day, it becomes the obvious route.

When you try to make a new trail, the first few steps feel slow and awkward because the ground is rough. The more you walk that new path, the easier it becomes. If you walk that trail enough times, the old trail fades and the new one becomes natural. That's the gist of plasticity. It's the way your brain reshapes itself through use and attention.

Plasticity matters in fast-changing environments because the brain loves efficiency. It leans on those existing pathways, which is why things like smoking and drinking are so hard to quit. It keeps you fast but it also locks you into old patterns. When the environment changes, efficient pathways can become blind spots. They help you feel like you're on stable ground even when you're wrong. Plasticity lets you build new pathways that respond to new information by getting you out of default mode. That way, you can make decisions based on the world you're actually in rather than the one you remember.

Three things support plasticity.

The first is novelty. The brain pays attention when something feels new. New information increases activity in the brain regions that support learning and pattern building. Novelty shakes the brain out of autopilot and even small doses help.

The second support is repetition because even small changes need repetition to create a new pathway. It'll be weak at first, but you'll build strength by using it consistently. Repetition packs more insulation around the connection which makes the signal faster. That is how habits form. Oddly enough, it's also how unlearning works because you're strengthening the new circuit and starving the old one.

You can see this in your daily behavior. The first time you change a routine, it feels forced, but by the fiftieth time, you do it without thinking.

The third support is emotional intensity. Emotion boosts everything, including learning (and unlearning), because it raises the chemicals that cement those new pathways. Think about some of your strongest memories, both good and bad. You remember those events because they have an emotional charge, which just goes to show that those signals create stronger physical marks inside the brain. This doesn't mean you need to have a run in with your old bully to form new habits, but we need to acknowledge that emotional relevance speeds up neural change. When something matters to you, the brain treats it as high priority so the pathway forms faster. On the other side of that coin, when your emotional intensity is low, your brain moves slower because it doesn't see a reason to change.

Now, underneath those three supports, there are biological mechanisms that make unlearning possible. One of them is synaptic pruning, which I already hinted at. The brain trims away weak or unused connections (Cafasso, 2018). It does this to save energy and improve efficiency. When a connection isn't reinforced through thought or behavior anymore, the brain lets it fade. That makes space for stronger pathways to take over. Pruning is the biological reason old habits lose their grip on you when you stop feeding them.

The last piece of our brain pie is long term potentiation (Shors, 1997). Connections get strengthened and really rooted through this process because it's based on repeated activation. In simple terms, every time you reinforce a new pattern, the signal becomes stronger. Long term potentiation practically guarantees that the new pathway becomes the default. Then, we've got long term depression, which acts like the counterpart to long term potentiation. That's what fades the old route.

What's the point of all of this?

Well, you've got some unlearning to do if you're going to be able to spot signals (and then actually do something with them).

Cognitive Rigidity and Its Effects

When you're trying to learn everything in this book and unlearn some things in the process, you're going to feel like you're about to doze off. These are high-level topics and no matter how much pizzazz I add to this book, there is only so much fun you can inject into data dissection. All of that to say that you're probably going to hit a patch of cognitive rigidity around about now because we're about halfway through a carefully chosen word count.

Cognitive rigidity is something that tends to show up when your brain stops updating its patterns. When you keep using the same mental route because it worked before, your brain starts feeling that the familiar path is smoother and, therefore, safer. It filters out new information, including the signals you need to spot, because it doesn't treat what might pass for emotional triviality as something worth adjusting for.

Routine reinforces the pattern (as well as successful case use) and that makes you resistant to new ideas. When the brain has locked into an old pattern, anything unfamiliar feels like a disruption because new ideas ask the brain to spend energy. They ask the brain to explore paths that don't match the model it has leaned on for years. That feels uncomfortable, so the brain pushes back. None of this is intentional. It's the natural behavior of a brain that doesn't want to rebuild a pathway unless the situation forces it to.

You can see this in the way people respond to change. A rigid mindset will almost always question the point of a new approach and defend the current system despite the fact that the evidence shows the cracks. That bias is going to explain away new signals because those signals feel inconvenient. When you see it happening, you can stop it from blocking your road to innovation.

This is the opposite of what you need to be doing if you're in a leadership position. A rigid leader expects the world to behave the same way it did when they first succeeded. They trust the methods that gave them status. They fall back on the same stories and the same decision logic. When a new idea shows up, the leader might see it as a distraction from 'real' work. They might think the team is overreacting and convince themselves that they're the only voice of reason when they're actually handcuffing the entire organization to an outdated map.

Organizations are always going to follow the mindset of the leaders and rigidity grows like a weed when the leaders stay fixed. Teams that start preferring predictability over curiosity are going nowhere, so make sure your culture isn't holding on to old playbooks even when the market shows a clear change. Repeat after me. Experimentation is not a threat. Your culture should reward it instead of avoiding it.

You have to think adaptively.

Expertise as a Barrier to Adaptation

Your expertise has grown over the years. You weren't born in a suit and tie. That is, unless you're Boss Baby, and if you don't get the reference, well, fair enough. The point is that you have to recognize that this is something that has grown with time and should, therefore, be allowed to continue evolving as more time passes. The problem is that the more experience you collect, the stronger your mental patterns become. When you know a space well, your brain responds to new information by comparing it to what you already believe. If the information doesn't match, you feel an urge to drop it. This is where blind spots are.

So, cognitive rigidity can come from expertise if you're not careful and I call this the expert's dilemma, which is based on the idea that the very thing that made you effective becomes the thing that slows you down.

Expertise shapes how you feel. You trust the old model because you won with it and likely built your name with it. You protect the old rules because the old rules gave you your wins. Change feels especially foreign to you because you're subconsciously defending your identity. Fresh signals also look like extra work and leaders with deep experience want efficient patterns and predictability. New ideas break that flow, so your first reaction might be irritation. The fact is that experience ages fast because markets change quickly and behavior goes in a different direction practically overnight. New tools show up before people finish learning the old ones, so your expertise doesn't update at the same speed. You might think that you're ahead because you have years' worth of knowledge, but old rules lose power.

The Adaptive Mindset

The answer to the problem is taking on an adaptive mindset, which is the opposite of cognitive rigidity. It's somewhat of a mental posture that gives leaders the ability to update their thinking as the world evolves around them. It might sound exhausting to constantly have to evolve with it, but that just comes with the territory of being a leader or being a marketer. That's been true even before AI came around. This means that you've got to let go of old maps and actively watch for new signals. Adjust when the evidence changes and treat your past successes as context instead of guarantees. This mindset might look simple on paper. You might be reading this and thinking that it's a no-brainer, but in reality, it requires a couple of traits that most people never practice in a deliberate way. The good news is that those traits can be trained.

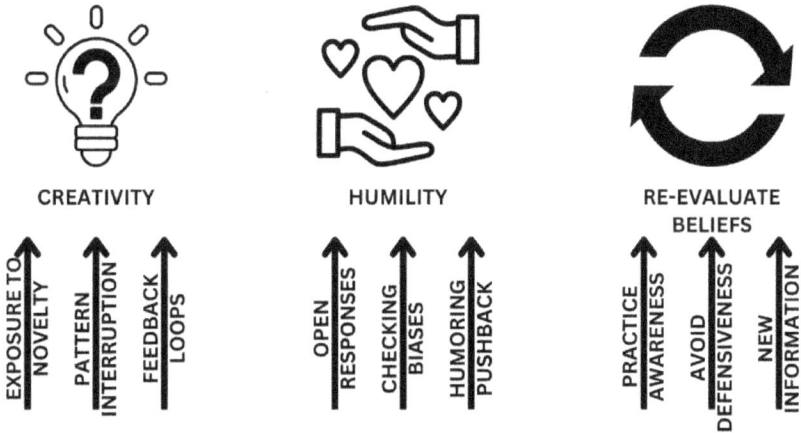

CREATIVITY

EXPOSURE TO NOVELTY   PATTERN INTERRUPTION   FEEDBACK LOOPS

HUMILITY

OPEN RESPONSES   CHECKING BIASES   HUMORING PUSHBACK

RE-EVALUATE BELIEFS

PRACTICE AWARENESS   AVOID DEFENSIVENESS   NEW INFORMATION

Curiosity is the first trait.

Curiosity keeps your brain open long enough for you to notice the signals that others ignore. Curious leaders constantly ask questions and that goes the same for the situations in which they think they know the answer. Being able to explore ideas that feel strange, including the idea that you might be wrong about something, is how you get to the point of being able to sit with possibilities that don't match your past experiences. You can use curiosity to shake off cognitive rigidity because it has a way of stopping your brain from locking into old patterns. This is the 'explore before you judge' path. If you want to get more curious in your daily life in general, pay attention to what feels new and to the tension around you. When something doesn't fit your mental model, analyze it to the nth degree.

The next trait is humility.

Humility is practical in this context because it's in the way that you actually engage with the world around you. You have to understand that the world is changing too fast for any one person to hold a complete picture. You're human, so accept it, and then create something that can at least attempt to fill the mental gap between certainty and assumption. It should be a net of sorts because it should still have space for new information to settle in. Practice saying, "I don't know." Try it. You can't possibly know everything, so it shouldn't be so outlandish of a thing to say. You're not saying that you're not willing to find out, though. That one sentence will keep your mind open and invite teams to share signals that they may have kept to themselves. In environments that have been shaped by speed, your access to new information is actually an adaptive advantage.

Last, we have a willingness to reevaluate beliefs.

This is the trait that ties the whole mindset together. It's easy to believe something because it has always been true, but it's much harder to check whether that belief still matches the present moment. Adaptive leaders do this constantly by returning to their assumptions and asking whether those assumptions truly reflect the world as it is right now. When they pick up on a mismatch, they change direction. It's not some personal failing, so don't beat yourself up over it. Think of it as maintenance or a way to keep your internal map updated so that your decisions stay sharp.

I know that it's not easy. In fact, I'd rank training your brain to stay open instead of being defensive as a real skill. The natural reaction to anything new is usually, "No thanks." I've said it before. The brain likes what it already knows because it takes less effort, so you have to train it past that point. When something pokes your point of view, ask what would need to be true for this thing to matter. Stop the knee-jerk reactions and try to bring in new input on purpose by listening to people outside your field. The more new things you feed your brain, the easier it becomes to move when the world moves and, eventually, move before it does.

Practice

At the end of the day, pick one assumption you felt like you relied on a little too heavily. Ask if it's still true. If it holds up, great. If it doesn't, update it instead of carrying it into tomorrow. I'll give you an example.

Assumption = People want cheaper options right now.

Curb it by looking for situations in which customers chose a more premium product even though the economy was sluggish. You might notice rising chatter online about "buying fewer but better."

Pattern interruption will help you with that. When you feel yourself falling into the same automatic response, switch tasks. Make it a conscious action of thinking about how you can interrupt the pattern and then actually do that. If you're struggling, ask for real input from people who don't gain anything by agreeing with you and then actually listen without trying to defend your point. If you hear the same thing from more than one person, you can take it seriously.

These are your first steps toward getting through bias and blindness.

Chapter 7 –
Bias, Blindness, and Breakthrough

Unlearning is only the first step when it comes to transforming your assumptions. Your next step is actually finding the blind spots that you aren't aware of. If you were aware of them, they wouldn't be called blind spots, so don't try to convince yourself that you know where they are.

I'll say it one more time. You're human.

You're going to have strategic blind spots and you're going to use two super simple tools to work around them. The fact is that if you don't address them, they'll eventually become your own undoing. You understand that to some degree because we've gone over how mindset affects the way leaders see the world. You've also been shown how curiosity, humility, and reevaluation keep your thinking flexible. When you've tackled that, you'll have cleared the mental space that you need for the next layer of strategy. Once your mind is open, you can finally see what most leaders miss.

Always remember that marketing teams fall behind when they rely on the same mental model that carried them through past wins. They keep using old assumptions even after the market starts transforming. The only solution is to transform along with it.

Your Strategic Blind Spots

I'm not going to tell you what your blind spots are because I'm not aware of your individual situation. So, what I'll do now is walk you through the one blind spot (which leads to many others) that I see so many leaders fall victim to.

Blind spots often come from an easy-to-make mistake that hardly anyone notices in the moment. You might look at the right signals but read them through the wrong clock. If you check the market on a short timeline when the change is on a long one or watch long term tension and expect it to pay off in the next quarter, you might have time as one of your blind spots. It's one of the worst to have because you end up expecting early signals to line up with your current goals. You might also treat fast noise like a permanent change and treat slow change like nothing at all. You're essentially choosing the wrong time frame when you look at the world.

What you need is a time horizon, which is the scale you use to judge what a signal means. If the horizon is too short, you'll miss signals that take time to form. If the horizon is too long, you'll miss small moves that need your immediate attention. If the signal is visible but your interpretation isn't all that clear, you'll likely have judgment that's based on the wrong clock.

There isn't one right or wrong way to gauge the time horizon, either, because both short- and long-term thinking can cause problems.

If you use your next quarter as the main lens, you're in short term mode and you might judge everything that you do through performance targets. When you're coming at the picture from that angle, you almost expect behavior to change fast and for the market to grab new ideas right away. You see an early signal and you want instant traction. When it doesn't happen, you decide the change isn't real and miss the point, which is that some of the biggest changes start slow because they need time to work their way into people's daily lives. Slow signals can lead to some of the most important movements, but if they fall outside of the short lens that you're used to, you'll brush them off.

This is how long-term movements get ignored. If you're prone to spotting a new pattern inside a small group, watching it for a few weeks, and then giving in because the growth feels flat, you're on shaky ground. It might not be the fad you think it is. Don't assume that the category will stay the same because the pattern might not be weak. It might be early and it just needs time as well as culture to catch up.

Long-term thinking comes with its own blind spot. This tends to happen when you extend your lens too far and expect every change to follow a big or slow arc. This leans on that experience bias we looked at in the previous chapter. If you expect categories to hold their shape for years and assume nothing major will happen in the near term, you will most definitely miss out on a short change. The worst thing to do is to see small moves in your space and brush them off because they look too tiny to matter. Have you seen how gargantuan a mustard tree can be in comparison to the seed it grows from? Keep that in mind because if you trust the long story more than the present moment and the market doesn't wait for long arcs, you'll be in trouble. Remember that the market moves the second people start moving, so a small spike can turn into something real when the conditions are ready. If your lens stays too wide, you won't see the spike until it has already turned into momentum.

Kodak went wrong with its long-term view and with its tendency to lean on past wins. In the early 2000s, small digital point-and-shoot cameras started getting tiny bits of traction. The numbers were small and the units were niche. The profit was also low, so Kodak brushed it off because the signals were too small to challenge a market that sold billions of dollars of film each year.

But those tiny spikes were real and they missed out on the short change because they were still too focused on the long game. As soon as early adopters started moving, the mass market followed faster than Kodak's long arc could explain. Digital adoption grew in a sharp curve, which means it spiked practically overnight. Teen girls all wanted digital cameras in pinks and purples. Before anyone knew it, film was a thing of the past.

Kodak couldn't accept the spike because its lens was too wide (Mui, 2012). There are tons of mega companies that have gone this way. Blockbuster is one of them, too, but it went the opposite route by seeing the early signals of change and mistaking them for weak signals because they were slow to develop. Blockbuster had an opportunity to buy out the company that would become its top competitor, but it didn't. Executives laughed them (Netflix co-founders Reed Hastings and Marc Randolph) out of the room (Satell, 2014).

You see, the wrong time horizon also creates blind spots in competitive analysis. If you judge your rivals on your own timeline, the picture gets distorted. What I mean by that is that if you expect every competitor to move at your speed, you'll miss the ones who are moving faster. If you assume their innovation cycle matches yours, you'll panic over moves that aren't real threats. From there, you're only going to push a signal into a timeline that doesn't fit and your read on it won't hold up. You can't be so hung up on your timeline that your marketing suffers, and trust me, it will suffer the most from this mistake. Messages built for the long term get tested on short term cycles. Short term spikes get projected into long term plans. It's just a mess waiting to happen.

If you want to reduce this blind spot, you need to match the horizon to the signal. If the signal comes from a fast cultural movement, you use a short arc. If it comes from a deep identity change, you use a long arc. If it's tied to tech adoption, you treat it as a mixed arc. If it shows up in behavior patterns, you use a steady arc. If it comes from product cycles, you use a seasonal arc. If you choose the horizon based on the signal instead of your planning cycle, the blind spot starts to fade. To do all of that accurately, it's best to have an inversion strategy.

Inversion Strategy

Inversion is a thinking tool that helps you notice what your normal viewpoint might be hiding. Most people look at a plan and ask how it can work. Inversion asks how it can fail. You ask what could break it and look for what you're assuming without proof. It sounds like incredibly negative thinking, which it can be if you approach it that way. The point here isn't to be negative, though. What you should be aiming for is a buffer of best-case and worst-case scenarios so that you can fall somewhere in the middle. When you do this, weak spots will show up that you couldn't see before.

Inversion works because every plan sits on assumptions that you rarely check. Some are solid assumptions but some only feel that way because, as I said before, you've used them for years. Some are just habits that nobody questions. When you invert an assumption, you test its real strength and you'll be able to see the real risk underneath it. That means that you can also spot new openings that were hidden before the flip. There are a couple of questions I typically ask to spark inversion thinking. Have a look and pick the ones that match your signal.

Inversion Questions

1. What if the opposite is true?
2. What if this belief works today but breaks tomorrow?
3. What if our strongest belief is our greatest weakness?
4. What if the customer no longer wants the reason we think they buy from us?
5. What if the risk we think is small is the risk that grows the fastest?
6. What if the category rules are wrong?

Inversion exposes a lot of hidden risks because it forces you to test the thing that you'd rather avoid. When you ask if the opposite might be true, you're pushed to name the conditions that would break the plan. When you ask if your strongest belief might be your weak spot, you find places where you're overconfident. When you ask if the category rules might be wrong, you open space for new positioning. It might feel like you're taking two steps back, but the answers are useful.

Inversion won't just reveal weak spots, which is somewhat of a relief. If you take the time to unpack the weak spots, it will also lead you to a ton of opportunities because the opposite angle can show you areas that no one else has claimed. A belief that feels locked from the front can look wide open from the back. I'll give you some more examples of this so that you can visualize exactly what I'm referring to.

Examples of Inversion in Action

1. A team believes customers buy a product for convenience. Inversion asks what happens if they are actually buying for identity. (Outcome is new messaging.)

2. A brand believes competitors are a threat that will play out in two years. Inversion asks what happens if the threat arrives in six months. (Resource planning is affected.)

3. A leader believes a trend is small. Inversion asks what happens if the small trend carries a sign of long-term change. (Changes how the team tracks signals.)

4. A company believes price is the barrier. Inversion asks what happens if emotional friction is the real barrier. (Opportunity for repositioning.)

Now you need a practical method that your team can use that is both short and repeatable.

Practical Inversion Framework

1. Write the assumption you are using to make your decision. Make it one sentence.

2. Flip the sentence. Write the version that contradicts it.

3. List the conditions that would make the flipped version true.

4. Check whether any current signals match those conditions.

5. Decide whether the flipped case shows risk or opportunity.

6. Record the change. If the flip exposes something, adjust the plan.

This framework gets you to a place where you're trying to see the full picture. Once you can do that, you can build a strategy that survives real conditions instead of imagined ones. There is one more area that can provide you with a breakthrough and I'll walk you through it now.

Reverse Mentorship

Reverse mentorship gives leaders a direct line into the parts of culture that move first. For the most part, leaders and marketers sit far from early signals. There's either an age or lifestyle gap that prevents them from seeing the pulse as it ignites. Sitting with people who are close in age and close in background might work from a professional standpoint, but it can put you out of touch with the metaphoric boots on the ground.

The long and short is that the future never starts in those rooms and I'm pretty confident in using an absolute with this. The edge networks that we looked at earlier are typically made up of younger people who live a lot closer to (or right within) emerging behavior. They feel the changes long before the numbers confirm it. Suffice it to say, you should be actively listening to those voices. Bringing them into your decision-making lets you see the world for what it is.

Younger voices are edge-positioned thinkers who have an advantage because their daily life keeps them close to cultural movement. They jump onto new platforms as they appear and move between digital identities with hardly any effort. It all comes naturally to them and they usually form part of these niche beta groups that get to "test" new tech without knowing that they're doing it. Everything that's new to them gains traction faster because they aren't carrying habits that have been built over decades like older generations have. Ultimately, they pick up on early friction because they live right at the center of the same forces that shape younger consumers just like them. When you listen to these voices, you get a lens that sees farther and faster. Their age and socio-economic positioning give them proximity to change.

Traditional hierarchy usually blocks that access because information gets pushed through so many layers that soften the message. Still, the two work hand in hand because people at the bottom see new behavior first and people at the top make decisions last. That said, we have to acknowledge that by the time a signal moves upward, it becomes safe and edited to match what the leader already believes. This is why leaders hear about new behavior long after it starts. That's especially true if you don't have an open organization that rewards creativity but instead rewards following structure. The top ends up living in a different world from the people who see the future forming.

Reverse mentorship cuts out the filter because it gives you a straight line from the edge to the top. You hear the raw signal instead of only seeing the cleaned-up version near the end of its life cycle. This is the early tension that I spoke of in the previous chapter. If you were wondering where or how you find it, this is it. You get it at the grassroots where it hasn't turned into numbers yet. That tension is gold because when you get it early, you make better calls.

To make reverse mentorship actually work, you need structure. You're not going to get insight through a casual chat with someone younger. You have to make it a real practice, so start off by picking someone who sits close to culture. This might be a young employee or a content creator. The industry isn't really an issue. In fact, it could be anyone from someone in tech to someone who's active in niche online spaces. The point is that they need to live where new behavior starts and be a curious person by nature. They also need to be honest and have the freedom to tell you the truth without feeling like they have to protect your feelings.

Ideally, there are a couple of boxes you'll want to tick off and you can have an assistant or someone from HR handle this process of whittling back options if you prefer.

Area	Question to Ask	
Cultural proximity	What online spaces or communities do you spend the most time in each week?	☐
Platform fluidity	How fast do you switch to new platforms or features when they show up?	☐
Early tool adoption	What new apps, tools, or features have you tried in the past month and what pulled you toward them?	☐
Honest feedback style	If you saw something in our work that felt off or outdated, how would you tell me?	☐
Emotional Read of a Culture	When something new starts to grow online, what tells you that people feel excited about it?	☐
Story pattern awareness	When you see a new format or meme, how do you know if it is just a joke or something that will stick?	☐
Real-time awareness	What changed this week in your feeds that felt different from last week?	☐
Curiosity in practice	What rabbit hole did you fall into recently and what did you learn from it?	☐
Social movement freedom	When your friend group shifts behavior, how do you notice it first?	☐
Ego-free communication	Do you feel comfortable disagreeing with people who outrank you and can you give me an example?	☐

You might have other areas that you want to cover, but these are the general ten that you should consider. When you've found your person, you can set a regular rhythm. You can't do this once and say you "get it." Culture moves too fast, so you need short weekly sessions where they tell you what they saw, including new content patterns, new conversation energy online, new products people are excited about, and changes in language. Look for sparks and listen for small patterns that keep showing up.

Now, you can streamline this by having a listening group of younger individuals in specific industries. You can buy insight from them or trade their insight for something that they might want, but if you're going to go that route, you're putting yourself at risk. We have to understand that human behavior is what it is and it can't be changed or controlled by an outside force. If they start feeling like the relationship or group is just a means to an end (some extra cash for themselves or something they want), they might just start looking for easy ways to provide insight without it actually meaning anything. If you systemize it with the same set of questions, you'll box them in and they'll start answering for the sake of it without their answers being really valid.

So, instead of trying to get the insight of a thousand people by over-systemizing this particular process, start off with one or two people and make a relationship out of it. You'll get a lot more real feedback over coffee every Tuesday than you will over an e-form.

You also need clear expectations. This young mentor shouldn't feel like they have to impress you. Their job is to show you what you can't see from where you sit. They need to feel safe pointing out strange behavior and calling out your blind spots. They need to feel safe saying when something feels old. Your job is to respond with curiosity and get your ego out of the way. If you snap or defend yourself, the whole thing dies. The signal won't reach you if the person delivering it feels like they have to protect your comfort as if you're a toddler.

The last step is documentation. Every session should leave you with a simple log of signals that will become part of your insight system. You can compare them to other signals by mapping them as I've shown you earlier. You can track which ones grow stronger over time and, bit by bit, you'll build a running picture of cultural movement that gets sharper with each round.

This is how you begin unlocking new patterns and get yourself further out of your expertise biases.

Part IV: <u>U</u>nlock New Patterns

(Rewire)

Chapter 8 –
The Mechanics of Momentum

This is when you can really start to unlock new patterns in your processes. You'll need to keep reminding yourself of the mindset that you've built up because we've been conditioned to think that growth follows a straight line. Life in itself doesn't move at a calm and predictable pace. Our world is becoming increasingly globalized, which means the butterfly effect truly exists. What happens in a warehouse in one of the major economic countries can impact every other country across the world. So, if you're relying on looking at your own environment (which might appear safe and predictable), you're already ten steps behind. That picture works for small, local, and linear growth, which all add the same amount in trajectory upswing over time. It feels easy to measure, and it is, but exponential growth behaves in a very different way. It multiplies instead of adding and it grows by stacking gains on top of gains. That stack makes the early phase look slow. Then the curve turns into a hockey stick and suddenly everything around you is changing.

We know about the quiet start because we've been over that together. What we need to pinpoint is the place where exponential change hides in the beginning. The fact is that you can spot early exponential indicators if you know what to look for when the data starts moving on the page.

The S-Curve of Adoption

The S-Curve helps you read how an innovation grows. You use it to see where a signal sits and where it is heading. The curve has four phases. These are emergence, acceleration, maturity, and saturation.

E EMERGENCE A ACCELERATION M MATURITY S SATURATION

Each phase has a pattern that you can track and when you understand those patterns, you can spot timing windows and decide where to place your resources. To build an S-Curve for a signal, start with a simple timeline. It can be monthly or quarterly. Then track the two numbers that are of the most importance, which are participation and pace. Participation shows you how many people are using the new idea. Pace shows the speed of the change. If you can't get clean numbers, track proxies. These would be the following.

- Search volume
- Social mentions
- Creator usage
- Community growth
- Conversation frequency
- Product trials.

These proxies tell you where the energy is moving.

In the emergence phase, which is what we've actually focused on in the previous chapters, the curve looks flat. The numbers rise but the rise is tiny, so you'll see small groups using the new idea and hints of emotional pull inside niche circles. You see early novelty. I've given you the means to decipher those stages, but when the acceleration begins and the rise starts to multiply, you need to be evolving with the data faster because you'll see more people adopting the idea each period. A practical way to confirm acceleration is to compare the last three periods. If each one grows faster than the one before, you've entered acceleration.

Maturity appears when the rise slows but stays high. Growth becomes a little more stable because the signal isn't a novelty anymore. You can confirm maturity by checking whether the last few periods show a rise that stays level. There should be no jumps or spikes.

Lastly, we have saturation, which happens when participation in the market flattens out. You'll see almost no rise and competitors will fill the space. You can confirm saturation when the number barely moves across several periods. The signal has reached its limit.

The most important skill is spotting the inflection point. This is the moment the curve changes shape. There are three you need to track. The jump from emergence to acceleration. The slowdown from acceleration to maturity. The flattening from maturity to saturation. You find the inflection point by watching how fast the number changes. When the speed shifts, the curve shifts.

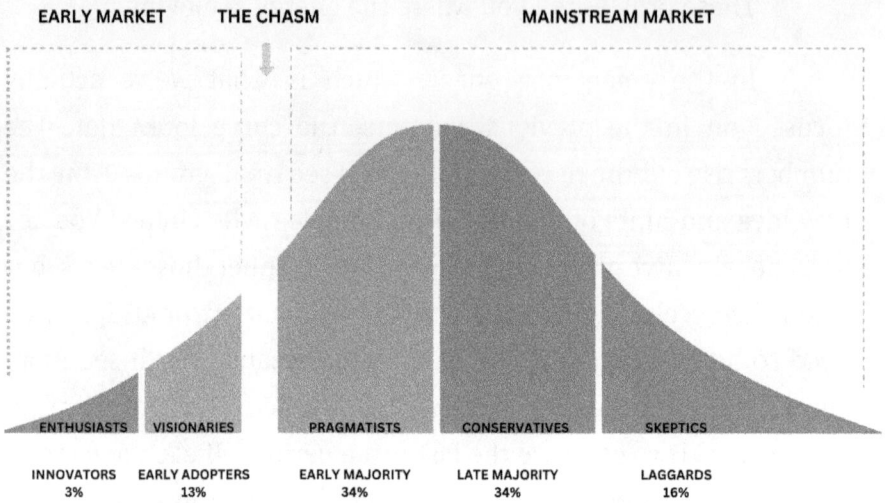

EARLY MARKET	THE CHASM		MAINSTREAM MARKET	
ENTHUSIASTS	VISIONARIES	PRAGMATISTS	CONSERVATIVES	SKEPTICS
INNOVATORS 3%	EARLY ADOPTERS 13%	EARLY MAJORITY 34%	LATE MAJORITY 34%	LAGGARDS 16%

(Glenn, 2009)

Here's what these shifts look like in real numbers. If weekly searches for a new tool rise by 5%, then 6%, then 12%, you've reached the jump. If growth climbs by 15% for a few periods and then drops to 6%, you've hit the slowdown. If the number rises by 1% or stays flat, you've reached saturation.

Once you know the phase, you know how to make your next move. In emergence, you monitor. In acceleration, you invest. In maturity, you differentiate. In saturation, you prepare for decline or look for the next curve. These moves keep you from wasting resources and chasing the white rabbit.

When you understand where you are on the curve, you can also get a little deeper into the process of your acceleration dynamics.

Acceleration Dynamics

Velocity tells you how fast a trend is moving and you measure it by tracking the rate of change across time. Volume tells you how big the trend is right now and where it's going next. The goal is to measure the slope instead of trying to measure the size. When the slope rises fast, you have acceleration. When the slope stays flat, you have stability. When the slope falls, you have a decline. You use velocity to see whether a small signal is gaining power or fading out.

The most basic way to measure velocity is to compare two time periods. As I said before, you'd pick a short window like one week or one month and track the number of mentions or searches or uses inside that period. Then, you'd compare it to the next period. That said, although I've given you the premise of these elements in the context of how they need to be rolled out, you'll also need to become a bit more elaborate by learning the formulae.

Velocity = (Current Period Value - Previous Period Value) / Previous Period Value

This gives you the percent change. A 10% rise per period is slow. A 30% rise is medium. A 50% or higher rise is strong. If the percent change grows each period, you have real acceleration. Another way to measure velocity is by watching how fast the trend moves across channels. If you see the same idea on TikTok, then see it on Reddit within days, then see it on YouTube within the same week, the velocity is strong. You can measure this by timing the jumps. There is a formula for this as well.

Channel Jump Velocity = Number of Channels Adopted / Number of Days Between First and Last Jump

A higher score means the trend is moving fast. A low score means it is stuck in one place. To unpack it more, search data also helps you track velocity. You can compare search volume by week. If the numbers double three weeks in a row, you have exponential velocity. You can calculate this with a simple growth multiplier.

Growth Multiplier = Current Week Searches / First Week Searches

A multiplier that is above two in a short window tells you that you've got fairly high speed. A multiplier above five tells you that you've got explosive speed. You can also measure creator adoption. Track how many creators mention or use the trend each period and use the same velocity formula. The creator signal is important because creators often move faster than brands. If creator velocity rises before consumer velocity rises, the trend is heating up. Also, make sure that you're tracking how fast reposts or likes and comments rise. You can measure this by tracking the average engagement per post during each period. If the engagement rate rises faster than the post count, the trend has real momentum. Again, you can use the same percent change formula.

A practical way to track velocity is to build a simple table with three base columns. List the trend. List the metric. List the weekly or monthly change. You can then add more columns in for tracking purposes. If the change keeps rising, the velocity is strong. If the change falls, the trend is cooling off. It's quite simple to get a clear snapshot when you use these tools and the formulae require a very basic understanding of math to work out.

Trend	Metric Tracked	Week 1	Week 2	Week 3	Velocity (W2 vs W1)	Velocity (W3 vs W2)	Direction
Short Video Editing App	Search Volume	12,000	15,600	21,840	+30%	+40%	Rising Fast
AI Meal Planner	TikTok Mentions	850	920	910	+8%	-1%	Cooling
Pocket Fitness Tracker	Creator Posts	42	78	160	+85%	+105%	Explosive
Quiet Luxury Fashion	Reddit Threads	19	22	21	+15%	-4%	Flattening
Digital Journaling Habit	YouTube Tutorials	1,300	1,495	1,680	+15%	+12%	Steady
Sleep Tech Wearables	Amazon Reviews Per Week	280	330	388	+18%	+17%	Strong Growth
Clean Energy DIY Kits	Google Searches	4,400	4,380	4,350	-0.4%	-0.7%	Declining

Your data plotting might look a little something like the table above. Now, if you're wondering where you can find the data and statistics, there are several tools and applications that you can tap into. You can track trend velocity with real tools that show change over time instead of total numbers. Search velocity is the easiest place to start because tools like Google Trends give you weekly and daily shifts. You can also use SEMrush or Ahrefs if you need deeper keyword data and monthly volume changes. Just remember that social velocity is just as important because a trend that spreads fast across channels will always outrun a trend that stays locked in one place. Tools like Brandwatch, Meltwater, and TikTok Creative Center help you track rising mentions and cross-platform jumps. These tools show the pace at which an idea is truly moving from one community to the next.

For creator velocity, platforms like CreatorIQ and Upfluence help you see how many creators pick up a topic each week. You can also measure community velocity through Reddit Metrics or

Subreddit Stats. These tools show how fast discussions grow inside niche groups. Engagement velocity is also easy to read with tools like SocialBlade or YouTube Analytics because they show how fast views and interactions rise over time. If you want to see product interest, Amazon Brand Analytics or Keepa will show you review velocity and price movement.

Now, when it comes to cross-channel jump velocity, BuzzSumo and Talkwalker help you read it. These tools will show you when an idea starts to appear on a new platform and how fast it spreads after the jump. Google Alerts can support all of this by tracking when new pages mention a keyword.

You don't need all the tools at once. It's just worth knowing them all (and doing your own homework on newer options as the years wane on) so that you have the ones that match the signal you want to measure when you want to measure it. Try to be consistent because the more you track the rate of change across channels, creators, communities, and search behavior, the easier it becomes to spot the trends that have real speed behind them.

Volume can fool you in the sense that a trend can look huge because it had attention in the past, but if the velocity is flat, it has no real future. A trend can look tiny because the volume is low, but if the velocity is climbing fast, you're looking at something with actual potential. When you're making early calls, you always choose velocity over volume because it tells you where the energy is. Volume only tells you where the energy used to be.

Once you know the velocity, you can decide what to do next. High velocity means you should test fast, medium velocity means watch it and get ready, low velocity means you should wait, and declining velocity means that it's time to move on. When you're in viable waters, you can see the stack.

Stacking & Convergence

When more than one advancement builds on another and they all start multiplying their impact, that's a stack. What you'll usually see is that one improvement increases the usefulness of the next, which creates this kind of compounding effect. Faster chips improve AI models, but then better models increase the value of automation tools, and then automation strengthens new workflows. Do you see how it amplifies with each advancement?

How do you spot it early? That's what we always want to know, right? Well, you can spot stacking when a series of small gains creates a jump that feels much larger than the individual parts.

All that said, we also need to pay attention to convergence because it's just as important but it works differently. Convergence is the point at which a couple of separate technologies or cultural forces blend together to form something new. This is the point where the lines between fields disappear and a new category starts to take its form and that new category becomes the baseline for how people operate. The move from phones and cameras to smartphones was convergence. Kodak at least had a little more than a hundred years before digital cameras became their "video killed the radio star moment." Digital cameras had about a decade (maybe less) before smartphones said "hold the phone!" Literally. The same goes for the change from fitness tech and personal identity to continuous health visibility.

You'll know that convergence is happening when the market stops treating the pieces as separate and starts treating them as one integrated system.

There are clear signs when stacking or convergence is creating a new baseline, the strongest of which is behavior change. You can confidently say that the baseline is forming when people adjust their habits to match the new pattern. If there's consistent language to go along with it and people describe the pattern the same way for months, the narrative is definitely settling in. You'll know that it's finally rooted when products or services start adopting the pattern across different categories.

Narrative collisions work at the story level and not at the structural level like you might think they do. A narrative collision happens when two rising story arcs meet and form a new meaning. Luxury plus sustainability created the eco luxury market. Authenticity plus AI created the idea of synthetic realness. Look for these collisions in language overlap first. You might notice that audiences use similar phrases across different topics and you might also notice cross following.

The huge upside in all of this is that narrative collisions give you category whitespace and you can shape your position before the market catches up.

AI-Assisted Momentum Detection

AI has changed the way you read momentum because it can catch patterns that you miss. Hardly any of us longer track trends by hand anymore because it's like guessing based on a handful of posts. You can use AI tools that scan millions of conversations and pick up tiny changes in language or behavior and these systems will help you see acceleration.

Several models already do this work. Large language models like GPT, Claude, Llama, and Gemini can pick up semantic shifts. They can scan past and present conversations to see how fast a new idea is gaining shape. They track the frequency of new terms and track the way those terms are used. They also track the emotional tone behind them. At the time of writing this, those LLMs make mistakes, so you have to stay on them. You have to make them confirm after every prompt that they've provided you with factual information and tease the thread. You also need to understand the constraints of the model that you're using and whether or not it's an enterprise or personal-use version.

Nonetheless, there are also tools that are built for acceleration detection. Tools like Signals, Blackbird, Delphi, and Feedly AI use machine learning to map weak signals and forecast where they're going. They track semantic clusters and audience movement while also measuring how fast a topic spreads from one community to another. These systems run on transformer models that excel at pattern recognition, which lets them pick up on micro shifts that would take a person months to see.

The only way to use these tools properly is to know how to read their outputs. Most of them show three things.

1. Rise rate
2. Spread rate
3. Saturation risk

Rise rate tells you how fast the conversation is growing. Spread rate tells you how fast the idea jumps across platforms. Saturation risk tells you if noise is growing faster than meaning. A high rise rate and a wide spread rate mean that you have real momentum to work with. A high saturation risk means that you're probably looking at nothing more than hype or a fad.

You can also ask a model to measure semantic drift, which will tell you how much the meaning of a topic changed in a set window. A small drift means the story is stable. A large drift means a new version of the story is forming. Large drift often happens right before acceleration, so take that as your early warning sign that the narrative is rewriting itself. From there, you can turn this into strategy by using acceleration models as triggers. A sharp rise in mention velocity means that it's the right time to test a position. A jump in cross platform spread means that you can (and should) invest more. A drop in drift means that you're at the perfect moment to root your story, while a spike in saturation risk means that it's time to pause and wait for a bit more of a clear view. You can apply the full framework with three checks.

Check velocity.

Check semantic drift.

Check spread.

If all three are rising, you move. If two rise and one stays flat, you prepare. If one rises and two fall, you wait. If all three fall, you exit. It's a simple system to implement that can become part of your scenario planning methods.

Part V: **R**ender Scenarios

(Cast)

Chapter 9 –
Designing Alternate Futures

When you've got finding signals, plotting their meaning, handling your assumptions, and unlocking new patterns under wraps, you can focus on becoming a little more inventive with strategic imagination. This helps you look and think beyond what's visible. It's not the same as creativity for entertainment or aesthetics. I like to think of it as a practical skill that helps you work with situations that don't really have a clear answer. Using imagination as a strategic tool means you get to practice exploring paths that don't exist yet, which has a way of preparing you for outcomes that feel unfamiliar or even a bit weird. It also widens your view because you stop relying only on the conditions in front of you.

This is a form of future-oriented thinking, which helps you move past fixed assumptions. Strategic imagination gives you range because you can picture several directions before you choose one. You're letting your thinking stretch forward and making space for possibilities that routine planning never really brings up. This protects you from getting tunnel vision because you're not locked into one prediction. On the contrary, you keep that field open, which gives you better options.

The added plus with this is that you'll also likely stop reacting to the problem right in front of you and you start shaping what comes next. Being able to act from a place of intention instead of pressure will help you change your mindset even further and uncertainty won't feel like such a threat. After all, it's just a raw material you can use.

You need to be mentally flexible to use your imagination. You need the ability to hold more than one outcome in your mind at the same time and be comfortable with the fact that two opposing pieces of information can coexist in varying degrees. You have to be willing to test different versions of the future without committing too early. You need to think in possibilities.

Thinking in Possibilities Instead of Probabilities

People tend to default to probability, but probability thinking focuses on what seems likely. It uses past data to forecast the next step and that works when the environment behaves in a predictable way. It falls apart when the landscape evolves or when new forces start to form. Probability pushes you to look at the options that already sit in front of you. It keeps your attention on the outcomes that look somewhat reasonable, but that habit is going to do you no favors because it shuts out potential.

Possibility thinking looks for what's possible. You open the field and invite outcomes that look strange or hard to explain. You let yourself walk along paths that don't match the past, which moves your thinking into a wider frame because you stop assuming the next chapter has to look like the last one (or even follow the same rules).

One of the biggest benefits of possibility thinking is the space it creates for low-probability-but-high-impact outcomes. These are the moments that change a market or break a pattern and they look unlikely until they happen. A leader who works from possibility will catch them early because they've already pictured the shape that those outcomes could take. This gives you a practical advantage because you can prepare while others wait for confirmation. You can also protect yourself from shock because you've already tested what the world might look like if the strange outcome wins.

When I'm trying to immerse myself in possibility thinking, there are two options that I typically lean on, and again, you don't have to use every tool in this book all at once. You use the tool that correlates with the stage of the cycle that you're in and based on what your objectives are. So, with that out of the way (again), the first tool I use in these cases is called the pivot lens. You look at your current plan and ask what would need to change for the plan to break. It shows you where your assumptions might fall apart. Once you see these points, you can create stronger options.

Here's an example. Let's say a fitness brand builds a new program based on constant gym access, but a sudden rise in at home training could break the plan. The pivot lens forces the team to ask what happens if gyms face lower foot traffic again.

Now, here's how to use it.

1. Write your plan at the top of a page.
2. List three events that could break the plan.
3. List what those events would change in your market.
4. Build one backup move for each event.

This keeps your strategy flexible and stops you from assuming the world will stay steady.

The next tool is the alternative pairing method. You pick two variables that look unrelated and imagine what happens if they move together. This method gives you new directions because it forces your mind out of straight line thinking. You probably won't use every scenario you create, but the exercise stretches your imagination and increases your options. An example here would be sleep data and dating apps. They seem unrelated enough. When you pair them, you get a scenario where people might match based on sleep rhythm or morning alertness. That idea sounds out there, but it reflects a culture that values health signals inside social identity.

Here's how you'd use alternative pairing.

1. Pick two variables from your industry.
2. Push them together even if they do not fit.
3. Ask what people gain or lose if both variables shift at the same time.
4. Pull out one or two directions that feel possible.

This gives you scenarios that your competitors will never reach if they rely on linear thinking. When you start creating scenarios, you can start mapping them.

Introduction to Scenario Mapping

Scenario mapping takes your new scenarios and gives them four parts each. These are drivers, uncertainties, consequences, and strategic options.

D DRIVERS **U** UNCERTAINTIES **C** CONSEQUENCES **S** STRATEGIC OPTIONS

Drivers are the forces that get the world going in a certain direction. These can be one of three things, which are cultural forces, market forces or technology forces. Uncertainties are the parts you cannot control. These are the factors that can swing in more than one direction. Consequences are the results you get when you combine a driver with an uncertainty. Strategic options are the moves you can make inside that future.

To build scenarios, you start by picking your drivers, so you'll usually choose just two or three forces that will affect your space in the next few years. You can also use tools like Feedly AI or Signals here. You can identify the big uncertainties when you have those key drivers. These are the parts that can change the entire shape of the driver. For example, automation can be a driver and regulation can be the uncertainty. When you put them together, you get a pattern that might make you ask what happens if automation gets faster and regulation tightens, for example.

The next step is to turn these patterns into divergent futures. You do this by building several versions of the world based on how each uncertainty moves. If an uncertainty moves in one direction, you get one future, and if it moves in the opposite direction, you get another. You create these futures so that your mind has range. In essence, your goal here is to build enough variation that you start spotting risks and openings you would normally miss.

You can make this concrete with tools like Miro or FigJam to build a basic scenario grid. You place your drivers on one axis and you place your uncertainties on the other axis. Then, you can fill each square with a version of the future. Now, if you want a bit more structure, you can use Kumu to map the way each factor connects, but don't forget that you can also use a large language model to help you list possible outcomes in each future to actually populate it. Models like GPT or Claude can point out second order effects (which we'll look at in Chapter 10) and help you see what might happen after the first obvious impact. They can also help you turn the futures into clear storylines. When you're done, your model might look something like the following one.

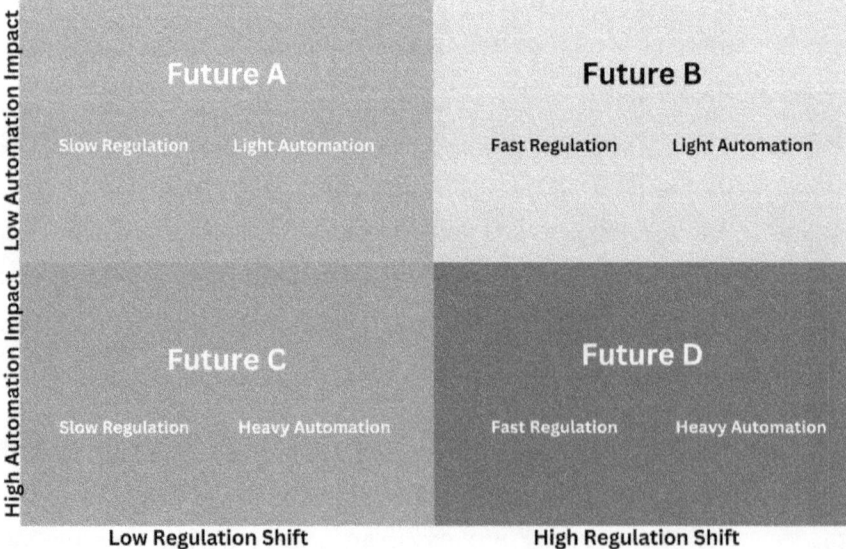

	Low Automation Impact		High Regulation Shift	
Future A			**Future B**	
Slow Regulation	Light Automation		Fast Regulation	Light Automation
Future C			**Future D**	
Slow Regulation	Heavy Automation		Fast Regulation	Heavy Automation

Low Automation Impact   High Automation Impact

Low Regulation Shift   High Regulation Shift

The example that I've just shared would be in relation to a strategic foresight question like the one I mentioned just before. You might want to know how different levels of automation and regulation could potentially shape the environment we operate in.

Once the futures are built, you can stress test your decisions by taking your current plan and dropping it into each future. When you do that, you can answer whether or not the plan survives pretty accurately just by looking at where the data lies. Marketing teams can test messages inside each future and leadership teams can test priorities while product teams can test features. Then you get to look at the forces that might shape those potential directions to see if the thought processes hold up.

Using the Futures Triangle

The Futures Triangle is what you'll use to understand the forces that shape a direction. That's because this is what you use to expose tension. Every future has three forces acting on it at the same time. There is a pull that draws people forward. There is a push that comes from the world right now. Lastly, there's a weight that keeps old patterns in place. You can see why a change feels stuck or why it moves fast when you map these forces. You can also gauge why it breaks open without warning.

The first force is the pull of the future. This sits at the top of the triangle. The pull comes from visions and desires, and it will show you the way people talk about what they want. The pull tells you what direction people hope to move toward even if they can't describe it clearly. When you map the pull, you list the things that attract people toward a new state. For example, you might list a desire for more creative freedom, a shift toward flexible work or interest in AI tools that help people build faster.

The next force is the push of the present. This sits on the right side of the triangle and comes from pressure that exists right now. It can come from rising costs, new tools, market tension or social expectations, to name a few. The push creates movement because it makes the present state harder to maintain, so focus on the conditions that are forcing change. Creator demand, platform shifts, new software, or new competition could form a part of your list. It's worth looking out for tech because the push often comes from technology that answers pressure inside communities.

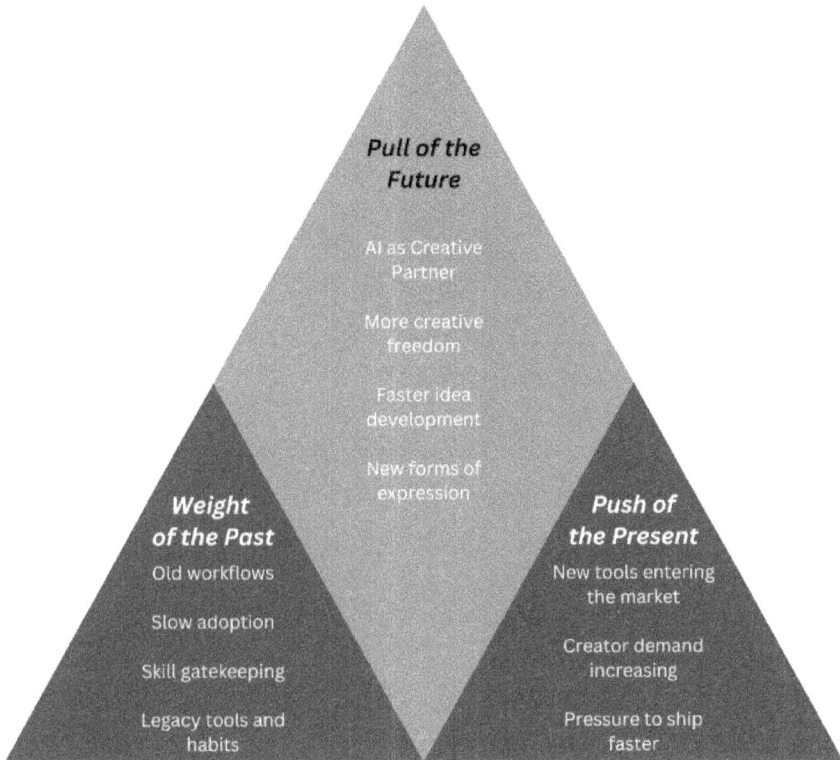

Pull of the Future
- AI as Creative Partner
- More creative freedom
- Faster idea development
- New forms of expression

Weight of the Past
- Old workflows
- Slow adoption
- Skill gatekeeping
- Legacy tools and habits

Push of the Present
- New tools entering the market
- Creator demand increasing
- Pressure to ship faster

(Inayatullah, 2008)

The last force is the weight of the past. This sits on the left side of the triangle and comes from the systems and habits that resist change. Look at rules or regulations that seem a little outdated or structures that have historically taken time to update. On this list, you should have the forces that slow change down. These could be anything from outdated workflows to old training systems, risk-averse teams or legacy thinking.

To apply the triangle, you pick a topic and build three lists that correlate to the topic before you place them in the triangle. As with all of the frameworks and tools I've given you so far, there's a way to read them. So, with this triangle, when the pull and push align and the weight is light, it means that the path is open. When the pull is weak but the push is strong, you probably need a stronger story. When the pull is strong but the weight is high, you'll likely need a better transition plan. Now, if the triangle doesn't give you deep enough insight, you can always use a wheel to stretch your data points instead.

Using the Futures Wheel

The futures wheel (Rogers, 1962) is a super simple way to map what a change might set in motion. To use it, you'll take one shift and stretch it into several layers so you can see what could follow, which helps you think forward instead of reacting to each step as it comes. The wheel works on one idea or a single event when you use it as a basis point and understand that it never ends at the first outcome. That first outcome pushes a second one and the second outcome pushes a third and so on. So, when you get down to it and map those layers on paper, you'll have a way clearer view of how the future might shape itself and transpire around the original change.

Building a futures wheel is just as easy as understanding and you'll start off by placing the signal in the center of a blank page and drawing a circle around it so your attention stays fixed. List the first effects around that circle and draw lines from the center to each one. These lines give you your first layer. Keep it to five to eight outcomes so you have a broad range without losing your focus. Then, you can look at each of those outcomes and ask yourself what might happen next. Place those answers in a wider ring and connect them with lines. This is your second layer. Repeat the same step for a third layer. Once you have three to four solid layers max, step back and study the shape.

A futures wheel lists outcomes but it also helps you see the consequences you didn't expect. You might find a hidden risk sitting two or three layers out or notice that several branches happen to meet in one point that shows pressure or opportunity. You can also find weak spots because if a negative effect appears in more than one branch, you have a vulnerability worth tracking. Essentially, that repetition is telling you where the future might push against you.

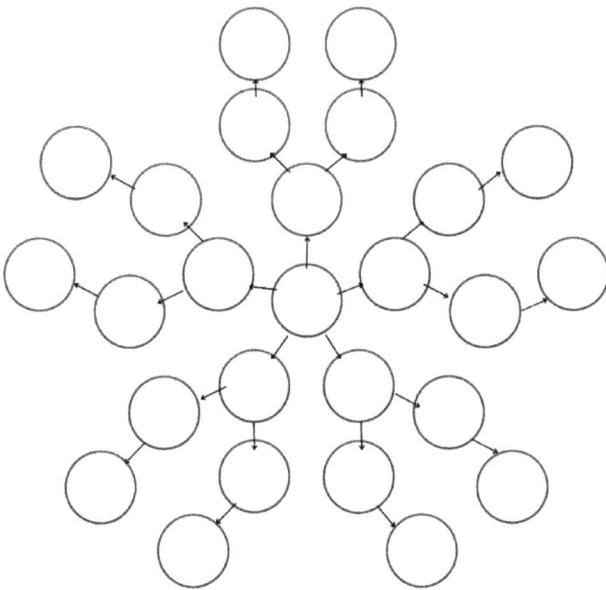

(Rogers, 1962)

You can use the blank wheel provided here in marketing. Place a new consumer behavior in the center and then map the first actions that might follow. You can then map the second and third layers. You'll possibly see a new demand pattern that doesn't exist yet or a story gap that you can shape. You might also see a positioning door that opens when the deeper effects find their respective places.

You can use the wheel for product work and place a new feature in the center. Map what users might do next or how support might change. I would also recommend mapping how expectations might change over a short period of time and looking at finding new hypothetical use cases. Really use this to your advantage because it'll help you take this imaginative forecasting and turn it into action.

Turning Strategic Imagination into Actions

This chapter has been all about designing alternate futures based on the signals and data that the earlier chapters had you capture. Strategic imagination is where you start to really boost that effort. However, you can't be imaginative for the fun of it. It only really matters in marketing when it helps you choose your next move. It's great to envision things, but imagination has more value when you take what you saw and turn it into concrete decisions. You can do this by moving your focal point from open exploration to clear direction by looking at the futures you built and looking at the patterns inside of them. I'd recommend that you specifically look for repeated areas because they'll show you where to focus.

The first step is to turn your scenarios into directional clarity and to do this, you need to pick a couple of the futures that you created. Place them side by side and look for the parts that appear in every future. These are called the durable themes. You can build on them because they hold their positions across many paths. Then look for the parts that show up in only one future. These are the outliers and they show you where risk is. When you compare the shared themes and the outliers, you start to see what you can act on today and what you should track before you commit. Now, don't disregard the outliers. You should actually track them because one global event could come along and change those themes.

In the next step, you're going to identify opportunity zones. An opportunity zone is a part of the future that has potential across more than one scenario. You can find these zones by looking for the areas where demand grows in more than one future as well as where constraints loosen in more than one future. These zones are generally strong because they don't depend on a single outcome, which ultimately means that you get to move early without taking a blind risk.

Miro or FigJam work well here, too, and they can help you to lay your futures next to each other. I would recommend that you use colored blocks to mark repeated elements and use another color to mark areas that feel promising. When you see a cluster of promising blocks across more than one future, you'll know that you have a zone worth exploring.

The next step is using the imaginative directions and scenarios that you created earlier to inform innovation. You should ideally be able to shape new ideas that fit specific zones, but you need to know the zones in the first place. You can design a product for a world where people want faster creation or a service for a world where people want more control, but you can only do that with zone awareness. If you're still struggling to see the right path, remember that you can always circle back to a previous step in the process and start testing more experiments or narratives. Just be sure that you're doing this because you can't see a clear path and not because you're gun-shy about pulling the proverbial trigger.

Use these tools for anything related to the upward trajectory of your business, organization, product or service. If you want to know where to invest or how to position yourself, this is how you do it.

Chapter 10 –
Consequences, Risks, and Openings

Strategic implications and the scenarios that you'll have built will help you understand what a present change might create later. An implication is a consequence that grows from a signal or a change you see right now. It's not the surface effect or the first reaction. Learning how to read implications means that fewer new changes come as an actual surprise. You'll eventually start to see how one move can trigger another and notice how small changes can stack until they turn into an entire industry transformation (birth or death).

Unlike an immediate effect that comes on fast, an implication takes on its life slowly. This slow formation creates the direction that comes after the noise, but you can only see it when you stop watching the reaction and start tracking the pattern. This is why I'm a firm believer that implication recognition is one of the core skills in foresight. You can't make sharp decisions without it. Thankfully, you can build this skill with simple habit adoption.

The first is one that I hinted at in Chapter 9.

Second-Order Thinking

First-order thinking stops at the first effect because it looks at the obvious outcome and puts on the brakes there. This looks like a new tool launching and people trying it. There's nothing more and nothing less to it. However, that's just the first step. They happen fast and everyone sees them. Second-order thinking looks at what the first effect will set in motion next and follows the chain. It shows you how a small change today can create a much bigger change later. When you take the time to practice second-order thinking, you'll automatically start looking for the next layer of consequence.

If you struggled to see how you would map the second ring of circles in the futures wheel, this is the type of thinking that will help you go back and amend those areas. I like to say that early strategic movers always think in cascades. They ask what the next step will be before the market reacts and they always expect a ripple. This helps them claim ground before anyone else notices what's coming. That's going to save you from trying to step in when saturation is already an issue.

I'll give you an example.

A few years ago, grocery stores started adding self-checkout lanes. The first-order effect was straightforward in the sense that lines moved faster because shoppers scanned their own items. The second-order effects weren't obvious at first, but after a while, there was a noticeable change in labor tasks. With fewer staff needed behind registers full time, they moved across the floor to handle exceptions or assist with the machines. Another second-order effect was a rise in small-scale losses because people scanned items (sometimes wilfully) incorrectly. Yes, they stole. The most expensive second-order effect was a change in store design because traffic moved differently. Of course, these things didn't happen on the first day of the initial change.

The point in all of this is that these second-order effects created both tension and opportunity at the same time. Some stores used the change to redesign layouts and place high-margin products near the new flow paths. Some stores added computer vision tools to cut losses and some added "scan as you go" carts. Most of these high-impact changes were second and even third order effects. They changed the path that the industry had been on and its rules.

Category Inversion & Industry Disruption

That's right where category inversion comes into the picture, which takes place when the logic of an industry gets turned on its head. A shift enters the market and the basic rules that once made sense start to feel outdated because the logic underneath them is changing. The category doesn't break because a better product shows up but because the old assumptions don't match how people think or what they value anymore. Then, the gap gets wide, and when it does, the category starts to split before a new one forms beside it. That new one grows fast and the old one becomes unstable.

You can see this in the move from ownership to access. People once paid to have things, but now they pay to use things. Value moved because convenience won over ownership. You can also see this in the move from product to experience. People once cared about features, but now they care more about how it feels to engage with the thing in question. These are value transformations in the flesh.

I'd start with the rules everyone in the category treats as normal and write them down. These rules show you how the category thinks the world works. Once you have them, flip each one and see what happens. If the rule says people want ownership, ask what the world looks like if they want access instead. If the rule says people buy for features, ask what the world looks like if they buy for identity. If the rule says people want workout plans, ask what the world looks like if they want real-time coaching instead. If the rule says people buy groceries weekly, ask what the world looks like if they buy meal-based kits on demand instead. You get the idea. These flips will show you where the pressure is, and after that, you'll go ahead and look for early signals that match the flips.

You can pick up structural change by watching for a few clear signs like those that we looked at in Chapters 1 to 3. Look for where the money and attention go. Use Google Trends to find rising interest, Brandwatch to see language change, and Signals and Feedly AI for early patterns that might still be gaining strength. If you want to follow the money, head to Crunchbase to see what investors have faith in right now.

Put these tools and processes and you'll start to see the real shape of the shift.

Part VI: <u>E</u>xecute & Embed

(Act & Train)

Chapter 11 –
Building a Future-Ready System

Now you get to execute and embed. That's the next logical step because foresight only works when it becomes part of how the team actually operates on a day-to-day basis. It shouldn't be looked at as something that one person on the team does on the side. This is part of the reason why I usually recommend weekly Signals Reviews, amongst other things that we'll look at in a moment, because they keep everyone tuned in to what's changing. You can meet at the same time every week with the same setup and you'll get the kind of feedback and results that you need.

This is the foundation that comes before cadence scanning or even foresight pods because it gets down to the bare bones, which are the signals that are still in an early phase. Ideally, everyone would bring two signals that they spotted in the past week. Before the meeting, everyone drops their signals into a shared sheet so that the team has one place to track everything. The meeting itself is ten minutes. That's it. The only objective is to walk through the signals and tag each one in one of three categories. These are rising, steady, or fading. Those tags will be your input for product calls, messaging shifts, planning cycles, and leadership updates.

Just this one ritual can change the entire pace of the company. It'll make the team more alert because everyone will be scanning. This then makes you all faster because you'll see micro movements before the data starts to build. When this becomes almost second nature, you'll want to focus a little more on cadences.

Scanning Cadences

Scanning has to become a part of your routine, but you need to start the process by deciding how often you actually want people looking outside the building. Different roles need different rhythms, so don't cram everyone into the same schedule. Yes, weekly scanning works best for the groups closest to culture, but that doesn't cover every department. I would suggest that you make these scanning drives more relevant for marketing, content, community, and research teams because these are the teams that sit right at the edge, anyway. By that, I mean they feel new energy first. A weekly rhythm helps them catch the fast patterns while they're still fresh.

You'll also need monthly and quarterly scanning, but monthly scanning fits leaders better because they need a wider view. They're not really the ones to be tracking micro shifts, so if that's you, this is your timeline. Look at the medium-range patterns, like category movement, platform changes, early business-model experiments, and changes in product behavior. You need more space because you have to think about speed and resources, so a monthly cadence will keep you in the know without forcing you into constant reaction mode.

That said, if you're going to do reverse mentorship or have it as part of one of your tactics, you might consider having your head of marketing handle it. That way, they get everything that they need from the edge networks (both within their team and outside of it if possible).

Quarterly scanning is where the long-term forces live, so things like tech curves, regulation, demographic changes, and shifts in values. You don't look at these every week because they don't move every week. It's that simple. However, they shape the ground you stand on, so a quarterly rhythm is a must to keep the company grounded in actual reality instead of basing long game moves on intuition alone.

At this stage, you should be creating the rhythms and once you've done that, you get to assign owners. Pick one person to own weekly cultural signals. Pick one to track industry movement. Pick one to follow tech changes. Pick one to map competitive activity. There needs to be a clear organogram for how this is all going to work. Now, them having ownership doesn't mean that they're doing everything alone. Not only is that going to silo you and the rest of the team out of that person's efforts, but it's also going to overload them. If that happens, you're one resignation away from not knowing where one of your routines leaves you. So, having one owner just means that they keep the flow tight and gather signals. The team adds what they find, but the owner filters through it.

The next step in the chain is finding a place to store everything because this is where most teams fall apart fast. When people send things at random without proper storage for the data, by the end of the month, half the signals are gone and the rest are floating around like digital lint. You need one shared board or server that works like a living radar. Again, you can build it in Notion, Miro or FigJam. They're really versatile, so that's why they've come up a lot, but the tool doesn't really matter. As long as you have a central home for everything, the board becomes relevant. It should be the place where every signal goes and you tag it with the date, the source, the category, and a quick sentence about why it's important. That board then becomes the memory of the team, which prevents those silos. Just make sure you're backing up your raw data because without it, the system will eventually collapse.

To keep everything practical, end each cadence with one action. Don't try to fully plan or brainstorm at that point. That will come as you build pods and cross-functional action. Ideally, this weekly team should eventually become your foresight pod.

Foresight Pods & Cross-Functional Insight Teams

The next link in the chain comes from these two teams and I'll tell you why you should consider creating one (obviously, depending on the scale of your operation). The short answer is that the book is called FUTURE for a reason. You're meant to be basing these moves almost entirely on forecasting and foresight is the precursor to that. The fact is that most company heads say they want foresight but we both know that what they really want is certainty, and if that's you, you need to stop because that mindset is exactly why nothing changes. The only real way to build foresight inside a team is to create a small group whose entire job is to stay close to signals and translate what they see. That's the role of a foresight pod. It's not a committee and it's not a task force. It's definitely not a group that meets once a quarter. It's a small set of people who see the world differently and who meet a couple of times a week to catch changes in the early market stage. (FYI, these are your enthusiasts and innovators.)

A pod is efficient because it mixes people who sit in different corners of the company. Marketing sees culture before everyone else. Product teams see friction before customers complain. Finance feels changes in spending. Customer support hears confusion in real time. When you put all of these people in one room, you get a real picture of the future without needing to wait for a report to be drafted and published.

Pick four to six people (depending on the size of your teams) and choose two who are great at pattern recognition, one who spends most of their day with customers, one who understands the numbers, and one or two who sit closest to the edge of cultural or tech behavior. They don't have to have a specific title or rank. If their skills match, they're in.

The pod can and should have the responsibility of clustering and framing. Remember, this is the process of grouping the signals that share the same energy. They should look for signals that touch the same audience behavior and point in the same direction even if they come from completely different spaces. The cluster forms and then the pod can map the implication just as I've shown you how to do throughout the book. This should be a priority for them because when you have a cluster that feels alive, you want to know what it unlocks, what it threatens, and what new choices it puts on the table. Keep it to one page by mapping the pressure, the opportunity, and the angle of move. This gives the whole team a quick read on where the tension is building.

The next step is proposing a trial move. That's the small test that runs inside two weeks like we looked at earlier. The team would do this to get a fast signal, so it should be something like a content test or a tiny offer. You could even consider a loose feature sketch.

The goal is to see how the market reacts. After that, the team can work on reporting so that everyone at a leadership level can see what the pod sees. After that, the entire broader team can follow through at different intervals. If the market pulls on the test, the pod will then hand the insight to the right team. That team decides whether to build, invest, shift positioning or wait. Bear in mind that your pod isn't responsible for building. They're your spotter, so their responsibility is to see the turn but not to execute along the full road.

The last action is rotation. Every six months, you'll need to rotate one or two members because fresh eyes keep the signal flows crystal clear. If the same people stay in the pod forever, the whole thing will go stale and the pod will become just another predictable meeting. To make sure you're tracking effectively, you'll need dashboards.

Dashboards & Visibility Systems

I've watched so many companies build dashboards that try to track everything and they end up making a mess. They add charts because the tool allows it and add widgets because they look nice. They cram in every metric that they can find, but then nobody checks it because it feels like a digital junk drawer. That's the fastest way to lose sight of what you're doing. Now, a foresight dashboard has one job, which is to show what's moving, how fast it's moving, and what that movement might mean for the next decision.

You can use any one of the tools that I've shared with you previously to create this dashboard and what you'll want to start with is the one question that makes the whole thing useful.

What does leadership need to see at a glance to make a call in the next seven days?

If the answer doesn't fit on one screen, the dashboard is wrong. You track four things, which are signals, convergence, velocity and customer friction. You track nothing else at this stage. Everything extra goes into another folder so you don't dilute the signal.

A simple layout works every time. One box for rising signals. One box for slowing signals. One box for narrative shifts. One box for convergence points. When a signal changes direction, the box updates. This is how you train the team to stay awake. The dashboard becomes something they check the way people check the weather. They know something's coming when the boxes change and they don't need a long explanation to understand what that means.

Whichever tool or framework you decide to go with, please remember that a dashboard only works when people can read it in seconds. Keep the design plain and the color coding simple. Designers love to add flair, I know, but flair can kill foresight. If people are scared to toy with it or add to it because they need an InDesign certificate to do it, you're sinking your own ship. Think of it like a hospital chart. Nothing can get in the way of the reader understanding it at a glance because they need to be able to make quick calls. If you don't know what hospital charts look like in terms of their plainness, give it a Google.

The whole point is that a dashboard will, without a shadow of a doubt, become useless if it's buried in some forgotten folder. It needs to be as clean as possible, but also needs to be visible and public inside the company. One of the best tips I've ever gotten is to place it inside the tools that people already use. If the team uses Slack, pin it to Slack. Pin it to Teams or Notion if that's what everyone prefers. It should be somewhere that you all can access without too much effort. There's already enough going on in the day to be hunting around for things.

Keep the loop small and predictable.

Chapter 12 –
The Future-Ready Leader

Emotional adaptability is a real advantage when markets move or your team gets stretched. Life is going to keep on doing what it does best and that means that plans are bound to break at one point or another. The one thing that is going to keep you stable is your ability to stay grounded when everything else feels off balance. So, who is this chapter genuinely for? Leader is such a loaded word and people tend to throw it around a lot, but I firmly believe that a leader is anyone who is responsible for a team (even a team of one or two). If you fit the bill, your emotional regulation is going to carry you and your team over the finish line. Now, the time when we need emotional regulation the most is usually the time when things are going to absolute trash. You don't really need to regulate on a good day. So, it's a skill that you'll need in order to stay on the level when you're dealing with stress spikes and signal clashes. Leaders who can regulate themselves make clearer calls because they aren't reacting from fear or trying to protect their ego.

The way I see it, intelligence helps you understand change, but you need adaptability to help you operate through it. You might feel like you've reached the peak when you can spot signals or trends, but even fewer people can stay functional inside that trend when it starts to become unpredictable and bends in a new direction. Remember, the world is never going to slow down on your account, so if you're a rigid thinker who needs perfect conditions to think straight, you will freeze when things change without warning. Adaptability keeps you moving because your internal state does not fall apart every time something unexpected happens. There are a few habits that you can work on at this point and they're all going to help you develop your adaptability before you can hone it for long-term leadership vision.

Leading with Long-Term Vision

The first of those habits is stress labeling. When something hits you, name the feeling out loud.

"I feel overwhelmed."

"I feel disappointed."

"I feel tense."

"I feel pressed for time now."

"I feel uncertain about the future of our product."

Calling it out reduces the punch because your brain stops treating it like a threat. Well, not immediately. Let's be realistic here. The point is that you get a bit of distance from the issue and, with distance, you can choose a response instead of letting the feeling run the moment and force you into one.

The next habit is super important for leaders and it's based on emotional rehearsal. To do this, you pick a scenario that normally triggers you and walk yourself through how you want to respond next time. You keep practicing it (to create a new neural pathway) and with enough repetition, you'll have given your brain a route before you need it. You're basically training yourself on how to react, which will leave you with a lot more flexibility in the moments that would have knocked you off balance before.

Then, you can focus on your long-term vision.

It sounds counterintuitive after all the talk about fast change and early signals, but vision is something different. Vision is not a signal, so you can play the long game with it. It's essentially the direction that your organization holds for the foreseeable future. It can be used to form your internal brand and culture dynamics, so it's quite important. When you strip it back, vision is just a psychological tool before it becomes anything strategic and I'll explain what that means.

To put it plainly, when people inside a company know where they're heading and what they're working toward, the uncertainty (which is just a part of marketing) loses its control over them. It doesn't have as sharp a set of teeth if your team members aren't all playing the guessing game about what the organization stands for and where they fit in. Give them a clear vision to calm the system and bring a bit of direction to a world that'll never slow down.

Now, let's try a quick exercise to help you build a long-term vision that you can stand on no matter what the environment is like. You're doing this to choose the future you want to help shape and to train yourself to return to it every week so that the world does not blow you off course.

Define the Future You Want to Shape

Sit somewhere quiet with a notebook. You're going to sketch the future in plain language. No jargon. No inspirational nonsense. Just clarity.

Step 1: Pick the customer promise you will never break.
"What is the one thing we stand for that still matters ten years from now?"

Write a single sentence.

Step 2: Choose the problem you'll still care about a decade from now.
"If everything else changes, what problem will still matter to the people we serve?"

Step 3: Choose the role you want your organization to play.
"When the category shifts, who do we want to be the reason for that shift?"

1 STEP 1: PICK THE CUSTOMER PROMISE YOU WILL NEVER BREAK.

2 STEP 2: CHOOSE THE PROBLEM YOU'LL STILL CARE ABOUT A DECADE FROM NOW.

3 STEP 3: CHOOSE THE ROLE YOU WANT YOUR ORGANIZATION TO PLAY.

When you have these three answers, write a plain-language paragraph that describes the future in a way that even a 15-year-old could understand.

Test Your Vision Against Your Week

Now that you have your vision, you need to test it against your week. After all, a vision is pointless if you never check your behavior against it. I'd recommend using this weekly routine to stop drift before it becomes a pattern.

Weekly Check-In Exercise

1. Look at your calendar from the past week.
2. Circle every decision, meeting, or priority call.
3. Next to each one, write:
   - Aligned
   - Not aligned
   - Unsure

If something isn't aligned, you'll need to fix it next week. This is technically your fourth step. If you are unsure, ask yourself if the future outcome that you want for the organization would approve the call you're about to make.

4   **STEP 4: DO A WEEKLY VISION TEST**

That one assessment does so much for exposing drift because it forces honesty out of you.

Embody the Vision So People Feel It

Teams don't follow you, so if your actions don't line up with the words on your website, your team is going to feel like you're a hypocrite. They're not going to do what you say and they'll probably start looking for more stable ground (now jobs) very soon. You've got to bring the vision into your behavior so that the team can read it without being told.

Step 5: Reflection Exercise

End each day with three short notes. I'll give you the questions that you'll need to answer to create them.

1. Where did I act in a way that supports the future I want?
2. Where did I wobble because of fear, pressure, or comfort?
3. What is one tiny behavior I can change tomorrow that points us back to the future?

5 **STEP 5: EMBODY WHAT YOU SAY**

You'll start to see patterns and really take note of the ways in which pressure tries to pull you off your direction. This daily reflection can train your body language and tone, but also your decisions, to match the future you keep describing to the rest of the team.

It's an embodiment practice that will make you more perceptive, which is where you want to be.

The Leader as a Perceptor

A leader becomes dangerous in the best possible way when they can read the room, the market and the culture without waiting for someone to brief them on it. I usually tell people that this part is personal because it's got nothing to do with dashboards or teams. It's entirely focused on your ability to pick up early movement before it becomes something everyone else finally sees. Now, I'm going to preface this by saying that you can't do it all. This is for the leaders of marketing teams specifically and not for organization heads or execs. As I said in previous chapters, your signal readers should be on your rotational teams. You should have reverse mentorship in place because you can't be everywhere at once. So when I'm talking about leaders in this section, I'm talking about the leaders or owners of those teams and subsets.

Back to the point at hand.

If you have strong perception, you can spot the difference between real movement and background filler. I've got an exercise for this, too, and it will make you more perceptive in every aspect of your life, which is ideal because we're all (well, most of us are) so unfocused thanks to our lifestyles.

The 5-Minute Sensory Sweep

Do this once a day in any environment. It could be in a coffee shop or even your workplace parking lot. Do it in your living room if you really want to. You're going to spend five minutes noticing things in a very deliberate way. You'll be picking up on the raw detail.

Set a timer for five minutes. Don't turn it into a meditation session or you might just zone out and not notice anything. This is training.

Step 1: Sight (1 minute)

Look at your environment as if you've never seen it and look for five things with shape or movement. I'll give you some examples.

- A person's shoes tapping at a different rhythm than the music.
- A building's paint fading in one corner but not the others.
- A shadow that looks slightly off because of a broken light.
- How the droplets of water form on your glass of ice water.
- The pattern that the blades of grass form around your feet.

Name each thing in your head. It will force your brain to register it, which you need in order to become more perceptive. The brain tends to want to gloss over these things because you see them all the time. So, it takes them for granted. Signals are also around you all the time, but your brain has learned to filter them out.

Step 2: Sound (1 minute)

Close your eyes for this one and try to identify four separate sounds. Try to find the ones buried under the loud sounds for extra points. I'll give you some more Examples.

- A fridge whirring or ticking under the general noise.
- A chair scraping lightly against tile.
- Someone muttering to their friend two tables away.
- The leaky faucet in the bathroom.

This teaches your brain to pull signals from beneath the obvious "noise", which is what perception actually is.

Step 3: Touch & Body Awareness (1 minute)

Move your attention to what your body feels.

- Where your clothes pinch or loosen.
- The weight of your phone in your pocket.
- The air on one side of your face being warmer than the other.
- The feeling of your hair tickling your ear.

You're teaching your brain to map micro-sensations instead of skipping them.

Step 4: Temperature & Smell (1 minute)

Notice the temperature around you and then identify two smells.

- Warm pockets of air near machines.
- The scent of metal from a railing.
- A faint cleaning chemical under a food smell.
- The smell of rain coming in cooling weather.

Step 5: Micro-Movements (1 minute)

Open your eyes again and look for the tiniest movements around you. I'll give you your last set of examples.

- A leaf blowing in the wind.
- Someone's foot moving or tapping every few seconds.
- A bag strap sliding down a chair.
- What your fingers look like when you turn a page.

This is going to sharpen your motion sensitivity. In real life, this is what picks up early changes. I can't stress enough just how much we take all of these things for granted and how dull our senses are truly becoming. When you get back to factory settings (like a kid), you're able to perceive so much more. It's a tightrope to walk because you need to be online to find the signals, but being chronically online is what's preventing you from being perceptive enough to find them. Trust me, this exercise will help with that. Then, you can become the architect.

Going From Operator to Architect

The move into architect territory starts when you decide you're in charge of the whole field and not just the tasks sitting in front of you. Operators focus on what's on today's list, but architects look at the whole space and tweak the way everything works. They influence the energy of the category and that's in part thanks to having an influence on how people talk and act. Throughout the book, I've said that this is where the big players end up. They don't predict futures or react to them. They create these futures because they're the architects.

Not everyone is going to get there and that's perfectly fine, but it's often better to aim for it and get as close to it as possible than not to try at all. There are tons of competitors who are going to be okay playing it small. Your effort alone is going to help you outpace them.

To get right down to it, this all starts with how you see your role. If you think of yourself as someone who just manages work, you'll stay stuck in the work. If you see yourself as someone who sets direction, your behavior will change and people will respond to the energy that you carry. Ultimately, your presence and the way in which you carry yourself both become a part of the job whether you mean it to or not.

Influence is the real power here because you can have the fancy title and still not move anything. People only follow you if they trust your read on what's happening, and that trust comes from quiet conviction. Architects build that feeling because they watch the deeper currents instead of chasing whatever blew up this week. That brings me to my next point. Momentum is where the big gap shows up. Operators wait until something is obvious, but architects look at early pressure and make moves before the target even knows that they want it. They just need enough clues to take the next step and they know that those clues are almost always buried in tension.

These are the people who get excited for customer feedback because they can pick a single sentence apart and mind-hack what the person is really complaining about.

Get excited about problems. That's my last bit of advice for you. If you can see past your own ego and start orchestrating what people want before they want it, you'll be on fire.

When all is said and done, "The best way to predict the future is to create it."

- Peter Drucker

In Closing

Foresight is a skill that you can build and you train it through repetition as well as by forcing yourself to look past whatever is sitting right in front of you. The more you work that muscle, the earlier you'll start spotting signals and the sooner you'll see patterns forming before the data catches up. The people who get ahead do it because they know how to read small moves before they turn into big ones.

FUTURE was built for exactly that, but keep in mind what I said right at the beginning of this book. I'm not asking you to memorize trends or lock yourself into fixed predictions. People take years to learn the principles and methodologies that I've shared with you. If you're lucky enough to have a large team, assign a tool or task to one member. That way, they can really focus on that one element without one person having to remember it all. The point is that my book is just a simple mental framework that you can use every day so you can read language changes and behavior cues. You should, by now, also be able to pin down emotional tone and early structural movement without getting lost in the nothingness of marketing noise.

That said, the real edge comes from perception, so you need at least one perceptive person on your team who is going to live on the edge (metaphorically, anyway). You'll need this because markets move fast and tech moves even faster. Anyone who waits for certainty will always show up after the upside has already been taken. You need the ability to pick up faint movement and you need enough confidence to act on early signals before they turn into mainstream talk. You'll get better at it.

To sign off, I'll say that the future pays the people who can see early. It also pays the people who keep their minds as flexible as the world that's moving around them. If you've learned one thing from all of this, I hope it's that you should treat foresight like a daily habit instead of a rare gift that only a few people are born with.

All the best out there.

Christina.

References

Aguilar, F. J. (1967). *Scanning the Business Environment.* New York : Macmillan.

Brewer, M. B. (1991). *The Social Self: On Being the Same and Different at the Same Time.* https://doi.org/10.1177/0146167291175001: Personality and Social Psychology Bulletin, 17(5), 475-482.

Cafasso, J. G. (2018). *What Is Synaptic Pruning.* Healthline: https://www.healthline.com/health/synaptic-pruning.

Christensen, C. M. (1997). *The Innovator's Dilemma: When New Technologies Cause.* Boston: Harvard Business School Press.

Feiner, L. (2019). *Facebook's Zuckerberg went before Congress a year ago — here's what has (and has not) changed since.* CNBC: https://www.cnbc.com/2019/04/09/facebooks-evolving-public-response-one-year-post-zuckerberg-testimony.html.

Giuntella O, H. K. (2021). *Lifestyle and mental health disruptions during COVID-19.* Proc Natl Acad Sci USA: https://pmc.ncbi.nlm.nih.gov/articles/PMC7936339/.

Glenn, J. C. (2009). *The Futures Wheel.* In J. C. Glenn & T. J. Gordon (Eds.),: The Millennium Project.

Haselton, T. &. (2019). *Here's a look at Apple's advertisement in Las Vegas ahead of CES.* CNBC: https://www.cnbc.com/video/2019/01/06/apples-ces-advertisement.html.

Inayatullah, S. (2008). *Six pillars: Futures thinking for transforming.* Foresight, 10(1), 4-21.

Mui, C. (2012). *How Kodak failed.* Forbes: https://www.forbes.com/sites/chunkamui/2012/01/18/how-kodak-failed/.

Prochazkova, E. &. (2017). *Connecting minds and sharing emotions through mimicry: A neurocognitive model of emotional contagion.* https://doi.org/10.1016/j.neubiorev.2017.05.013: Neuroscience and biobehavioral reviews, 80, 99–114.

Rienzi, G. (2024). *The science behind why we see faces in nature.* Hub: Johns Hopkins Magazine: https://hub.jhu.edu/magazine/2024/winter/pareidolia-faces-in-nature/.

Rogers, E. M. (1962). *Diffusion of Innovations.* New York: Free Press of Glencoe.

Satell, G. (2014). *A look back at why Blockbuster really failed and why it didn't have to.* Forbes: https://www.forbes.com/sites/gregsatell/2014/09/05/a-look-back-at-why-blockbuster-really-failed-and-why-it-didnt-have-to/.

Shors, T. J. (1997). *Long-term potentiation: what's learning got to do with it?* Behavioral and brain sciences, 20(4), 597–655: https://pubmed.ncbi.nlm.nih.gov/10097007/.

Wang JY, W. Q. (2023). *GLP-1 receptor agonists for the treatment of obesity: Role as a promising approach.* Front Endocrinol (Lausanne): https://pmc.ncbi.nlm.nih.gov/articles/PMC9945324/.